FREE Study Skills D...

Dear Customer,

Thank you for your purchase from Mometrix! We consider it an honor and a privilege that you have purchased our product and we want to ensure your satisfaction.

As a way of showing our appreciation and to help us better serve you, we have developed a Study Skills DVD that we would like to give you for FREE. This DVD covers our *best practices* for getting ready for your exam, from how to use our study materials to how to best prepare for the day of the test.

All that we ask is that you email us with feedback that would describe your experience so far with our product. Good, bad, or indifferent, we want to know what you think!

To get your FREE Study Skills DVD, email freedvd@mometrix.com with *FREE STUDY SKILLS DVD* in the subject line and the following information in the body of the email:

- The name of the product you purchased.
- Your product rating on a scale of 1-5, with 5 being the highest rating.
- Your feedback. It can be long, short, or anything in between. We just want to know your impressions and experience so far with our product. (Good feedback might include how our study material met your needs and ways we might be able to make it even better. You could highlight features that you found helpful or features that you think we should add.)
- Your full name and shipping address where you would like us to send your free DVD.

If you have any questions or concerns, please don't hesitate to contact me directly.

Thanks again!

Sincerely,

Jay Willis
Vice President
jay.willis@mometrix.com
1-800-673-8175

CHST
Exam
SECRETS

Study Guide
Your Key to Exam Success

Copyright © 2019 by Mometrix Media LLC

All rights reserved. This product, or parts thereof, may not be reproduced, stored in a retrieval system, or transmitted in any form or by any means—electronic, mechanical, photocopy, recording, scanning, or other—except for brief quotations in critical reviews or articles, without the prior written permission of the publisher.

Written and edited by the Mometrix Safety Certification Test Team

Printed in the United States of America

This paper meets the requirements of ANSI/NISO Z39.48-1992 (Permanence of Paper).

Mometrix offers volume discount pricing to institutions. For more information or a price quote, please contact our sales department at sales@mometrix.com or 888-248-1219.

Mometrix Media LLC is not affiliated with or endorsed by any official testing organization. All organizational and test names are trademarks of their respective owners.

Paperback
ISBN 13: 978-1-60971-350-8
ISBN 10: 1-60971-350-8

Ebook
ISBN 13: 978-1-62120-420-6
ISBN 10: 1-62120-420-0

Dear Future Exam Success Story

First of all, **THANK YOU** for purchasing Mometrix study materials!

Second, congratulations! You are one of the few determined test-takers who are committed to doing whatever it takes to excel on your exam. **You have come to the right place.** We developed these study materials with one goal in mind: to deliver you the information you need in a format that's concise and easy to use.

In addition to optimizing your guide for the content of the test, we've outlined our recommended steps for breaking down the preparation process into small, attainable goals so you can make sure you stay on track.

We've also analyzed the entire test-taking process, identifying the most common pitfalls and showing how you can overcome them and be ready for any curveball the test throws you.

Standardized testing is one of the biggest obstacles on your road to success, which only increases the importance of doing well in the high-pressure, high-stakes environment of test day. Your results on this test could have a significant impact on your future, and this guide provides the information and practical advice to help you achieve your full potential on test day.

Your success is our success

We would love to hear from you! If you would like to share the story of your exam success or if you have any questions or comments in regard to our products, please contact us at **800-673-8175** or **support@mometrix.com**.

Thanks again for your business and we wish you continued success!

Sincerely,
The Mometrix Test Preparation Team

Need more help? Check out our flashcards at: http://MometrixFlashcards.com/CHST

Table of Contents

Introduction	1
Secret Key #1 – Plan Big, Study Small	2
Secret Key #2 – Make Your Studying Count	3
Secret Key #3 – Practice the Right Way	4
Secret Key #4 – Pace Yourself	6
Secret Key #5 – Have a Plan for Guessing	7
Test-Taking Strategies	10
Hazard Identification and Control	15
Emergency Preparedness and Fire Prevention	77
Safety Program Development and Implementation	83
Leadership, Communication, and Training	92
CHST Practice Test	98
Answer Key and Explanations	130
How to Overcome Test Anxiety	150
Causes of Test Anxiety	150
Elements of Test Anxiety	151
Effects of Test Anxiety	151
Physical Steps for Beating Test Anxiety	152
Mental Steps for Beating Test Anxiety	153
Study Strategy	154
Test Tips	156
Important Qualification	157
How to Overcome Your Fear of Math	158
False Beliefs	159
Math Strategies	161
Teaching Tips	163
Self-Check	164
Thank You	165
Additional Bonus Material	166

Introduction

Thank you for purchasing this resource! You have made the choice to prepare yourself for a test that could have a huge impact on your future, and this guide is designed to help you be fully ready for test day. Obviously, it's important to have a solid understanding of the test material, but you also need to be prepared for the unique environment and stressors of the test, so that you can perform to the best of your abilities.

For this purpose, the first section that appears in this guide is the **Secret Keys**. We've devoted countless hours to meticulously researching what works and what doesn't, and we've boiled down our findings to the five most impactful steps you can take to improve your performance on the test. We start at the beginning with study planning and move through the preparation process, all the way to the testing strategies that will help you get the most out of what you know when you're finally sitting in front of the test.

We recommend that you start preparing for your test as far in advance as possible. However, if you've bought this guide as a last-minute study resource and only have a few days before your test, we recommend that you skip over the first two Secret Keys since they address a long-term study plan.

If you struggle with **test anxiety**, we strongly encourage you to check out our recommendations for how you can overcome it. Test anxiety is a formidable foe, but it can be beaten, and we want to make sure you have the tools you need to defeat it.

Secret Key #1 – Plan Big, Study Small

There's a lot riding on your performance. If you want to ace this test, you're going to need to keep your skills sharp and the material fresh in your mind. You need a plan that lets you review everything you need to know while still fitting in your schedule. We'll break this strategy down into three categories.

Information Organization

Start with the information you already have: the official test outline. From this, you can make a complete list of all the concepts you need to cover before the test. Organize these concepts into groups that can be studied together, and create a list of any related vocabulary you need to learn so you can brush up on any difficult terms. You'll want to keep this vocabulary list handy once you actually start studying since you may need to add to it along the way.

Time Management

Once you have your set of study concepts, decide how to spread them out over the time you have left before the test. Break your study plan into small, clear goals so you have a manageable task for each day and know exactly what you're doing. Then just focus on one small step at a time. When you manage your time this way, you don't need to spend hours at a time studying. Studying a small block of content for a short period each day helps you retain information better and avoid stressing over how much you have left to do. You can relax knowing that you have a plan to cover everything in time. In order for this strategy to be effective though, you have to start studying early and stick to your schedule. Avoid the exhaustion and futility that comes from last-minute cramming!

Study Environment

The environment you study in has a big impact on your learning. Studying in a coffee shop, while probably more enjoyable, is not likely to be as fruitful as studying in a quiet room. It's important to keep distractions to a minimum. You're only planning to study for a short block of time, so make the most of it. Don't pause to check your phone or get up to find a snack. It's also important to **avoid multitasking**. Research has consistently shown that multitasking will make your studying dramatically less effective. Your study area should also be comfortable and well-lit so you don't have the distraction of straining your eyes or sitting on an uncomfortable chair.

The time of day you study is also important. You want to be rested and alert. Don't wait until just before bedtime. Study when you'll be most likely to comprehend and remember. Even better, if you know what time of day your test will be, set that time aside for study. That way your brain will be used to working on that subject at that specific time and you'll have a better chance of recalling information.

Finally, it can be helpful to team up with others who are studying for the same test. Your actual studying should be done in as isolated an environment as possible, but the work of organizing the information and setting up the study plan can be divided up. In between study sessions, you can discuss with your teammates the concepts that you're all studying and quiz each other on the details. Just be sure that your teammates are as serious about the test as you are. If you find that your study time is being replaced with social time, you might need to find a new team.

Secret Key #2 – Make Your Studying Count

You're devoting a lot of time and effort to preparing for this test, so you want to be absolutely certain it will pay off. This means doing more than just reading the content and hoping you can remember it on test day. It's important to make every minute of study count. There are two main areas you can focus on to make your studying count:

Retention

It doesn't matter how much time you study if you can't remember the material. You need to make sure you are retaining the concepts. To check your retention of the information you're learning, try recalling it at later times with minimal prompting. Try carrying around flashcards and glance at one or two from time to time or ask a friend who's also studying for the test to quiz you.

To enhance your retention, look for ways to put the information into practice so that you can apply it rather than simply recalling it. If you're using the information in practical ways, it will be much easier to remember. Similarly, it helps to solidify a concept in your mind if you're not only reading it to yourself but also explaining it to someone else. Ask a friend to let you teach them about a concept you're a little shaky on (or speak aloud to an imaginary audience if necessary). As you try to summarize, define, give examples, and answer your friend's questions, you'll understand the concepts better and they will stay with you longer. Finally, step back for a big picture view and ask yourself how each piece of information fits with the whole subject. When you link the different concepts together and see them working together as a whole, it's easier to remember the individual components.

Finally, practice showing your work on any multi-step problems, even if you're just studying. Writing out each step you take to solve a problem will help solidify the process in your mind, and you'll be more likely to remember it during the test.

Modality

Modality simply refers to the means or method by which you study. Choosing a study modality that fits your own individual learning style is crucial. No two people learn best in exactly the same way, so it's important to know your strengths and use them to your advantage.

For example, if you learn best by visualization, focus on visualizing a concept in your mind and draw an image or a diagram. Try color-coding your notes, illustrating them, or creating symbols that will trigger your mind to recall a learned concept. If you learn best by hearing or discussing information, find a study partner who learns the same way or read aloud to yourself. Think about how to put the information in your own words. Imagine that you are giving a lecture on the topic and record yourself so you can listen to it later.

For any learning style, flashcards can be helpful. Organize the information so you can take advantage of spare moments to review. Underline key words or phrases. Use different colors for different categories. Mnemonic devices (such as creating a short list in which every item starts with the same letter) can also help with retention. Find what works best for you and use it to store the information in your mind most effectively and easily.

Secret Key #3 – Practice the Right Way

Your success on test day depends not only on how many hours you put into preparing, but also on whether you prepared the right way. It's good to check along the way to see if your studying is paying off. One of the most effective ways to do this is by taking practice tests to evaluate your progress. Practice tests are useful because they show exactly where you need to improve. Every time you take a practice test, pay special attention to these three groups of questions:

- The questions you got wrong
- The questions you had to guess on, even if you guessed right
- The questions you found difficult or slow to work through

This will show you exactly what your weak areas are, and where you need to devote more study time. Ask yourself why each of these questions gave you trouble. Was it because you didn't understand the material? Was it because you didn't remember the vocabulary? Do you need more repetitions on this type of question to build speed and confidence? Dig into those questions and figure out how you can strengthen your weak areas as you go back to review the material.

Additionally, many practice tests have a section explaining the answer choices. It can be tempting to read the explanation and think that you now have a good understanding of the concept. However, an explanation likely only covers part of the question's broader context. Even if the explanation makes sense, **go back and investigate** every concept related to the question until you're positive you have a thorough understanding.

As you go along, keep in mind that the practice test is just that: practice. Memorizing these questions and answers will not be very helpful on the actual test because it is unlikely to have any of the same exact questions. If you only know the right answers to the sample questions, you won't be prepared for the real thing. **Study the concepts** until you understand them fully, and then you'll be able to answer any question that shows up on the test.

It's important to wait on the practice tests until you're ready. If you take a test on your first day of study, you may be overwhelmed by the amount of material covered and how much you need to learn. Work up to it gradually.

On test day, you'll need to be prepared for answering questions, managing your time, and using the test-taking strategies you've learned. It's a lot to balance, like a mental marathon that will have a big impact on your future. Like training for a marathon, you'll need to start slowly and work your way up. When test day arrives, you'll be ready.

Start with the strategies you've read in the first two Secret Keys—plan your course and study in the way that works best for you. If you have time, consider using multiple study resources to get different approaches to the same concepts. It can be helpful to see difficult concepts from more than one angle. Then find a good source for practice tests. Many times, the test website will suggest potential study resources or provide sample tests.

Practice Test Strategy

If you're able to find at least three practice tests, we recommend this strategy:

Untimed and Open-Book Practice

Take the first test with no time constraints and with your notes and study guide handy. Take your time and focus on applying the strategies you've learned.

Timed and Open-Book Practice

Take the second practice test open-book as well, but set a timer and practice pacing yourself to finish in time.

Timed and Closed-Book Practice

Take any other practice tests as if it were test day. Set a timer and put away your study materials. Sit at a table or desk in a quiet room, imagine yourself at the testing center, and answer questions as quickly and accurately as possible.

Keep repeating timed and closed-book tests on a regular basis until you run out of practice tests or it's time for the actual test. Your mind will be ready for the schedule and stress of test day, and you'll be able to focus on recalling the material you've learned.

Secret Key #4 – Pace Yourself

Once you're fully prepared for the material on the test, your biggest challenge on test day will be managing your time. Just knowing that the clock is ticking can make you panic even if you have plenty of time left. Work on pacing yourself so you can build confidence against the time constraints of the exam. Pacing is a difficult skill to master, especially in a high-pressure environment, so **practice is vital**.

Set time expectations for your pace based on how much time is available. For example, if a section has 60 questions and the time limit is 30 minutes, you know you have to average 30 seconds or less per question in order to answer them all. Although 30 seconds is the hard limit, set 25 seconds per question as your goal, so you reserve extra time to spend on harder questions. When you budget extra time for the harder questions, you no longer have any reason to stress when those questions take longer to answer.

Don't let this time expectation distract you from working through the test at a calm, steady pace, but keep it in mind so you don't spend too much time on any one question. Recognize that taking extra time on one question you don't understand may keep you from answering two that you do understand later in the test. If your time limit for a question is up and you're still not sure of the answer, mark it and move on, and come back to it later if the time and the test format allow. If the testing format doesn't allow you to return to earlier questions, just make an educated guess; then put it out of your mind and move on.

On the easier questions, be careful not to rush. It may seem wise to hurry through them so you have more time for the challenging ones, but it's not worth missing one if you know the concept and just didn't take the time to read the question fully. Work efficiently but make sure you understand the question and have looked at all of the answer choices, since more than one may seem right at first.

Even if you're paying attention to the time, you may find yourself a little behind at some point. You should speed up to get back on track, but do so wisely. Don't panic; just take a few seconds less on each question until you're caught up. Don't guess without thinking, but do look through the answer choices and eliminate any you know are wrong. If you can get down to two choices, it is often worthwhile to guess from those. Once you've chosen an answer, move on and don't dwell on any that you skipped or had to hurry through. If a question was taking too long, chances are it was one of the harder ones, so you weren't as likely to get it right anyway.

On the other hand, if you find yourself getting ahead of schedule, it may be beneficial to slow down a little. The more quickly you work, the more likely you are to make a careless mistake that will affect your score. You've budgeted time for each question, so don't be afraid to spend that time. Practice an efficient but careful pace to get the most out of the time you have.

Secret Key #5 – Have a Plan for Guessing

When you're taking the test, you may find yourself stuck on a question. Some of the answer choices seem better than others, but you don't see the one answer choice that is obviously correct. What do you do?

The scenario described above is very common, yet most test takers have not effectively prepared for it. Developing and practicing a plan for guessing may be one of the single most effective uses of your time as you get ready for the exam.

In developing your plan for guessing, there are three questions to address:

- When should you start the guessing process?
- How should you narrow down the choices?
- Which answer should you choose?

When to Start the Guessing Process

Unless your plan for guessing is to select C every time (which, despite its merits, is not what we recommend), you need to leave yourself enough time to apply your answer elimination strategies. Since you have a limited amount of time for each question, that means that if you're going to give yourself the best shot at guessing correctly, you have to decide quickly whether or not you will guess.

Of course, the best-case scenario is that you don't have to guess at all, so first, see if you can answer the question based on your knowledge of the subject and basic reasoning skills. Focus on the key words in the question and try to jog your memory of related topics. Give yourself a chance to bring the knowledge to mind, but once you realize that you don't have (or you can't access) the knowledge you need to answer the question, it's time to start the guessing process.

It's almost always better to start the guessing process too early than too late. It only takes a few seconds to remember something and answer the question from knowledge. Carefully eliminating wrong answer choices takes longer. Plus, going through the process of eliminating answer choices can actually help jog your memory.

Summary: Start the guessing process as soon as you decide that you can't answer the question based on your knowledge.

How to Narrow Down the Choices

The next chapter in this book (**Test-Taking Strategies**) includes a wide range of strategies for how to approach questions and how to look for answer choices to eliminate. You will definitely want to read those carefully, practice them, and figure out which ones work best for you. Here though, we're going to address a mindset rather than a particular strategy.

Your chances of guessing an answer correctly depend on how many options you are choosing from.

How many choices you have	How likely you are to guess correctly
5	20%
4	25%
3	33%
2	50%
1	100%

You can see from this chart just how valuable it is to be able to eliminate incorrect answers and make an educated guess, but there are two things that many test takers do that cause them to miss out on the benefits of guessing:

- Accidentally eliminating the correct answer
- Selecting an answer based on an impression

We'll look at the first one here, and the second one in the next section.

To avoid accidentally eliminating the correct answer, we recommend a thought exercise called **the $5 challenge**. In this challenge, you only eliminate an answer choice from contention if you are willing to bet $5 on it being wrong. Why $5? Five dollars is a small but not insignificant amount of money. It's an amount you could afford to lose but wouldn't want to throw away. And while losing $5 once might not hurt too much, doing it twenty times will set you back $100. In the same way, each small decision you make—eliminating a choice here, guessing on a question there—won't by itself impact your score very much, but when you put them all together, they can make a big difference. By holding each answer choice elimination decision to a higher standard, you can reduce the risk of accidentally eliminating the correct answer.

The $5 challenge can also be applied in a positive sense: If you are willing to bet $5 that an answer choice *is* correct, go ahead and mark it as correct.

Summary: Only eliminate an answer choice if you are willing to bet $5 that it is wrong.

Which Answer to Choose

You're taking the test. You've run into a hard question and decided you'll have to guess. You've eliminated all the answer choices you're willing to bet $5 on. Now you have to pick an answer. Why do we even need to talk about this? Why can't you just pick whichever one you feel like when the time comes?

The answer to these questions is that if you don't come into the test with a plan, you'll rely on your impression to select an answer choice, and if you do that, you risk falling into a trap. The test writers know that everyone who takes their test will be guessing on some of the questions, so they intentionally write wrong answer choices to seem plausible. You still have to pick an answer though, and if the wrong answer choices are designed to look right, how can you ever be sure that you're not falling for their trap? The best solution we've found to this dilemma is to take the decision out of your hands entirely. Here is the process we recommend:

Once you've eliminated any choices that you are confident (willing to bet $5) are wrong, select the first remaining choice as your answer.

Whether you choose to select the first remaining choice, the second, or the last, the important thing is that you use some preselected standard. Using this approach guarantees that you will not be enticed into selecting an answer choice that looks right, because you are not basing your decision on how the answer choices look.

This is not meant to make you question your knowledge. Instead, it is to help you recognize the difference between your knowledge and your impressions. There's a huge difference between thinking an answer is right because of what you know, and thinking an answer is right because it looks or sounds like it should be right.

Summary: To ensure that your selection is appropriately random, make a predetermined selection from among all answer choices you have not eliminated.

Test-Taking Strategies

This section contains a list of test-taking strategies that you may find helpful as you work through the test. By taking what you know and applying logical thought, you can maximize your chances of answering any question correctly!

It is very important to realize that every question is different and every person is different: no single strategy will work on every question, and no single strategy will work for every person. That's why we've included all of them here, so you can try them out and determine which ones work best for different types of questions and which ones work best for you.

Question Strategies

READ CAREFULLY

Read the question and answer choices carefully. Don't miss the question because you misread the terms. You have plenty of time to read each question thoroughly and make sure you understand what is being asked. Yet a happy medium must be attained, so don't waste too much time. You must read carefully, but efficiently.

CONTEXTUAL CLUES

Look for contextual clues. If the question includes a word you are not familiar with, look at the immediate context for some indication of what the word might mean. Contextual clues can often give you all the information you need to decipher the meaning of an unfamiliar word. Even if you can't determine the meaning, you may be able to narrow down the possibilities enough to make a solid guess at the answer to the question.

PREFIXES

If you're having trouble with a word in the question or answer choices, try dissecting it. Take advantage of every clue that the word might include. Prefixes and suffixes can be a huge help. Usually they allow you to determine a basic meaning. Pre- means before, post- means after, pro - is positive, de- is negative. From prefixes and suffixes, you can get an idea of the general meaning of the word and try to put it into context.

HEDGE WORDS

Watch out for critical hedge words, such as *likely, may, can, sometimes, often, almost, mostly, usually, generally, rarely*, and *sometimes*. Question writers insert these hedge phrases to cover every possibility. Often an answer choice will be wrong simply because it leaves no room for exception. Be on guard for answer choices that have definitive words such as *exactly* and *always*.

SWITCHBACK WORDS

Stay alert for *switchbacks*. These are the words and phrases frequently used to alert you to shifts in thought. The most common switchback words are *but, although*, and *however*. Others include *nevertheless, on the other hand, even though, while, in spite of, despite, regardless of*. Switchback words are important to catch because they can change the direction of the question or an answer choice.

FACE VALUE

When in doubt, use common sense. Accept the situation in the problem at face value. Don't read too much into it. These problems will not require you to make wild assumptions. If you have to go beyond creativity and warp time or space in order to have an answer choice fit the question, then you should move on and consider the other answer choices. These are normal problems rooted in reality. The applicable relationship or explanation may not be readily apparent, but it is there for you to figure out. Use your common sense to interpret anything that isn't clear.

Answer Choice Strategies

ANSWER SELECTION

The most thorough way to pick an answer choice is to identify and eliminate wrong answers until only one is left, then confirm it is the correct answer. Sometimes an answer choice may immediately seem right, but be careful. The test writers will usually put more than one reasonable answer choice on each question, so take a second to read all of them and make sure that the other choices are not equally obvious. As long as you have time left, it is better to read every answer choice than to pick the first one that looks right without checking the others.

ANSWER CHOICE FAMILIES

An answer choice family consists of two (in rare cases, three) answer choices that are very similar in construction and cannot all be true at the same time. If you see two answer choices that are direct opposites or parallels, one of them is usually the correct answer. For instance, if one answer choice says that quantity x increases and another either says that quantity x decreases (opposite) or says that quantity y increases (parallel), then those answer choices would fall into the same family. An answer choice that doesn't match the construction of the answer choice family is more likely to be incorrect. Most questions will not have answer choice families, but when they do appear, you should be prepared to recognize them.

ELIMINATE ANSWERS

Eliminate answer choices as soon as you realize they are wrong, but make sure you consider all possibilities. If you are eliminating answer choices and realize that the last one you are left with is also wrong, don't panic. Start over and consider each choice again. There may be something you missed the first time that you will realize on the second pass.

AVOID FACT TRAPS

Don't be distracted by an answer choice that is factually true but doesn't answer the question. You are looking for the choice that answers the question. Stay focused on what the question is asking for so you don't accidentally pick an answer that is true but incorrect. Always go back to the question and make sure the answer choice you've selected actually answers the question and is not merely a true statement.

EXTREME STATEMENTS

In general, you should avoid answers that put forth extreme actions as standard practice or proclaim controversial ideas as established fact. An answer choice that states the "process should be used in certain situations, if…" is much more likely to be correct than one that states the "process should be discontinued completely." The first is a calm rational statement and doesn't even make a definitive, uncompromising stance, using a hedge word *if* to provide wiggle room, whereas the second choice is a radical idea and far more extreme.

BENCHMARK

As you read through the answer choices and you come across one that seems to answer the question well, mentally select that answer choice. This is not your final answer, but it's the one that will help you evaluate the other answer choices. The one that you selected is your benchmark or standard for judging each of the other answer choices. Every other answer choice must be compared to your benchmark. That choice is correct until proven otherwise by another answer choice beating it. If you find a better answer, then that one becomes your new benchmark. Once you've decided that no other choice answers the question as well as your benchmark, you have your final answer.

PREDICT THE ANSWER

Before you even start looking at the answer choices, it is often best to try to predict the answer. When you come up with the answer on your own, it is easier to avoid distractions and traps because you will know exactly what to look for. The right answer choice is unlikely to be word-for-word what you came up with, but it should be a close match. Even if you are confident that you have the right answer, you should still take the time to read each option before moving on.

General Strategies

TOUGH QUESTIONS

If you are stumped on a problem or it appears too hard or too difficult, don't waste time. Move on! Remember though, if you can quickly check for obviously incorrect answer choices, your chances of guessing correctly are greatly improved. Before you completely give up, at least try to knock out a couple of possible answers. Eliminate what you can and then guess at the remaining answer choices before moving on.

CHECK YOUR WORK

Since you will probably not know every term listed and the answer to every question, it is important that you get credit for the ones that you do know. Don't miss any questions through careless mistakes. If at all possible, try to take a second to look back over your answer selection and make sure you've selected the correct answer choice and haven't made a costly careless mistake (such as marking an answer choice that you didn't mean to mark). This quick double check should more than pay for itself in caught mistakes for the time it costs.

PACE YOURSELF

It's easy to be overwhelmed when you're looking at a page full of questions; your mind is confused and full of random thoughts, and the clock is ticking down faster than you would like. Calm down and maintain the pace that you have set for yourself. Especially as you get down to the last few minutes of the test, don't let the small numbers on the clock make you panic. As long as you are on track by monitoring your pace, you are guaranteed to have time for each question.

DON'T RUSH

It is very easy to make errors when you are in a hurry. Maintaining a fast pace in answering questions is pointless if it makes you miss questions that you would have gotten right otherwise. Test writers like to include distracting information and wrong answers that seem right. Taking a little extra time to avoid careless mistakes can make all the difference in your test score. Find a pace that allows you to be confident in the answers that you select.

KEEP MOVING

Panicking will not help you pass the test, so do your best to stay calm and keep moving. Taking deep breaths and going through the answer elimination steps you practiced can help to break through a stress barrier and keep your pace.

Final Notes

The combination of a solid foundation of content knowledge and the confidence that comes from practicing your plan for applying that knowledge is the key to maximizing your performance on test day. As your foundation of content knowledge is built up and strengthened, you'll find that the strategies included in this chapter become more and more effective in helping you quickly sift through the distractions and traps of the test to isolate the correct answer.

Now it's time to move on to the test content chapters of this book, but be sure to keep your goal in mind. As you read, think about how you will be able to apply this information on the test. If you've already seen sample questions for the test and you have an idea of the question format and style, try to come up with questions of your own that you can answer based on what you're reading. This will give you valuable practice applying your knowledge in the same ways you can expect to on test day.

Good luck and good studying!

Hazard Identification and Control

COMMON HAZARDS AND CONTROLS ASSOCIATED WITH HOT WORK

Hot work refers to work with hot metal such as brazing, welding, soldering, cutting with a torch, and drilling or grinding that potentially create a fire hazard. Employees engaged in hot work must be trained in the hazards posed by the work and how to mitigate them. A hot work program must include assessment of industrial hygiene hazards (exposure to metal dusts and fumes), assessment of noise hazards, assessment of proper personal protective equipment (PPE; eye protection, proper gloves, and respiratory protection if required) and must include an assessment of fire hazards posed by the hot work.

HOT WORK PERMIT SYSTEMS

A hot work permit ensures that safety precautions have been taken for welding or torch cutting activities. The permit is designed as a job aid to check that the work area has been prepared for flying sparks by removing **flammable debris** in a radius around the work area. The work permit should prompt a check that the **fire sprinkler protection system** is in place and functional, and that fire extinguishers of the correct type are staged and ready. The permit should include a provision to test the work environment to ensure there is no **explosive atmosphere**. The permit should document that the employees have received the required training and have the proper PPE available. The permit should include a provision to have a **fire watch** in place for at least thirty minutes after the hot work has been completed, and for the area to be periodically monitored for at least six hours after the work is completed.

SUPERVISOR'S DUTIES FOR HOT WORK

Supervisors are responsible for planning the work and ensuring that the hot work permit system is implemented. The use of the hot work permit checklist is a means of ensuring the proper safety precautions are taken and planning is completed prior to the hot work beginning. The supervisor is responsible for ensuring employees have the proper PPE on hand. Supervisors are responsible for providing the proper manpower and personnel to use a designated fire watch employee if necessary. The supervisor is also responsible for ensuring that employees have received the required safety and operational training to conduct the hot work.

FIRE WATCH

Proper planning for hot work or welding includes having a person to act as "fire watch." This is necessary for any activities that have a fire risk. The fire watch person should be trained in the duties and should understand the requirements of the **hot work permit**. The person should be stationed in the area and remain alert for sparks or embers emitted during the work. The person must be prepared to use a fire extinguisher and to alert other workers and authorities in the event of an emergency. Fire watch duties must extend after the hot work is finished until there is absolutely no chance that any embers remain that could cause a fire. If the fire watch must leave the work area at any time during work, work must be stopped until he or she returns.

WELDING PROJECT HAZARDS AND CONTROLS

There are several hazards to consider when planning a welding project. The safe use of the compressed gases used must be considered. The types of metals to be welded should be considered to determine whether hazardous metal fumes or dusts will be generated that will require respiratory protection (and specialized equipment that will fit under a welding helmet). The eye hazard posed by exposure to the wavelengths of light emitted by the welding process must be considered and the appropriate welding glasses or face shield obtained. The risk of fire and sparks must be considered, and appropriate protective clothing should be worn.

Welders will be exposed to sparks when welding. If the welder is wearing the wrong clothing or gloves, these sparks can **ignite** the clothing and burn the worker. Proper protection for the eyes and face include a **welding helmet** with face shield or goggles. The work uniform should be of **heavy-duty cotton** rather than synthetic materials because synthetic materials such as polyester have a lower ignition temperature. Clothing can also be manufactured from flame-resistant materials such as Nomex®. For additional protection, welders can wear a leather **apron** that will protect the chest and abdominal area.

COMMON ELECTRICAL HAZARDS AND CONTROLS

The Occupational Safety and Health Administration (OSHA) Electrical Safety Standard (29 CFR 1910.303) considers that guarding and personal protective equipment (electrical safety gloves or other measures) must be taken when employees are exposed to live electrical wires greater than 50 volts. Many companies have adopted more stringent requirements that require protection if exposure levels are greater than 30 volts. Electrical safety gloves must be worn, and untrained and unauthorized employees must stay at least 10 feet from the energized parts. There must be a means of triggering an emergency stop available to the employee, and there must be personnel on site who are trained in first aid measures for electric shock if employees will be exposed to electricity above these threshold levels.

ELECTRICAL SHOCK

An electrical shock occurs when the human body unintentionally becomes part of an electrical circuit. Electric shock occurs when:

- A person's body becomes part of the **conduction circuit or path** followed by electrical energy.
- A person becomes part of the circuit by holding both **positive and negative wires** of an electrical circuit.
- An otherwise closed electrical circuit loses or has a break in its **insulation**. Any person holding the line then becomes part of the circuit.
- Standing in or touching another **conductor carrying a current**. This is often caused by persons standing in water and making contact with electrical lines.

IMPACT OF ELECTRICAL CURRENT FLOWING THROUGH THE HUMAN BODY

Depending on the strength of an electrical current flowing through the human body, the impact can be minimal to fatal. The factors that determine the **severity of impact** are:

- The actual amount of current flowing through the body.
- The path that the electric current takes as it passes through the body.
- The length of contact the human body has with the sources of the electrical current.
- The frequency of the electrical current flowing through the body.

Assuming that the length of contact with an electrical source is at one second, an exposure of up to **5 milliamperes of current** can be experienced with only minor discomfort. As the current increases beyond 5 milliamperes for one second exposures, the exposures become gradually more dangerous.

ENERGY CONTROL SYSTEMS

Workers are often injured when performing service or maintenance operations on equipment operated by electric or other motorized power. The equipment being serviced must be secured against possible accidents by an energy control program. The employer's energy control program should train employees in safe procedures for servicing machines and equipment which might be subject to unexpected startup or operation. Tagout and Lockout devices are most often used to prevent this type of accident.

- **Tagout** refers to the placement of a tag or warning label which informs workers that a machine cannot be operated until the tagout device is removed by the health and safety officer.
- A **Lockout Device** is any mechanical method of preventing a machine from releasing energy or starting up operations while it is being serviced or maintained. Simple locks and keys may be sufficient. Other situations may require blocks, flanges, and bolted slip blinds.

TRAINING REQUIREMENTS

The Occupational Safety and Health Administration (OSHA) lockout/tagout regulation lists specific training requirements for employees both initially and on an annual basis. The training must cover the types of hazardous energy that can/should be controlled (not just electrical energy), what are authorized and affected employees, the specific lockout procedures for equipment they will be locking out, the difference between lockout and tagout, why tagout alone is not an approved control of hazardous energy, and the process to commence safe start-up of equipment after lockout.

AUTHORIZED AND AFFECTED EMPLOYEES

An authorized employee under the lockout/tagout regulations is an employee who has been thoroughly trained in the reasons for lockout and the methods to lock out equipment and has been issued a lock to use. The lock should be personalized by color coding, and/or by the use of personalized tags to notify others who is responsible for a particular lockout. An affected employee is one that has a general awareness of what lockout/tagout is and works near machinery that will be locked out from time to time. However, these employees are not responsible themselves for performing the lockout or for following the steps to safe start-up of equipment after lockout.

GFCI

A Ground Fault Circuit Interrupter (GFCI) protects against the most common type of electrical shock, a ground fault. A ground fault occurs when an electrical current finds an alternate 'path of least resistance' to the ground (or earth) than the ground wire. A person receives a shock when this path of least resistance is through their body. GFCIs are commonly used and required in wet areas such as bathrooms or kitchens, since the presence of water increases the chance that electrical current will experience a ground fault. Water is very conductive and provides an easy path for electrical current. The GFCI works by comparing the current that exits the unit to the current returning to it; if a significant difference is detected, it immediately cuts off the electricity flow through the circuit and protects from the risk of electrical shock.

Employer Responsibilities Regarding the Use of Ground Fault Circuit Interrupters

Employers have a responsibility to provide appropriate ground fault circuit interrupters for power tools and electrical equipment at construction sites.

- While the program is the direct responsibility of the employer, the employer may delegate the **implementation** of it to a "competent person."
- **120 Volt receptacle single phase outlets** on construction sites which operate at 15 and 20 amps, and which are not part of permanent building wiring, must be equipped with GFCI for protection of workers.
- The provision requiring adequate **ground and circuit interruption** must apply to all equipment available for use by workers. It applies to all cord sets and receptacles not part of the permanent building structure.
- The employer must maintain a written **description of the provisions** of the electrical safety program on site. The written documentation of safety provisions must be available for inspection by any regulatory or safety official and by employers who are affected by the provisions of it.

Exceptions to GFCI Requirements for Electrical Generators

There are **exceptions** to the OSHA provisions regarding GFCI circuit protection for non-permanent electrical circuits. When electricity is being delivered to small tools via portable generators, the frame of the generator may in itself serve as the grounding electrode mechanism for the electric circuit supplied by the generator. This is true when:

- The **portable generator** is the supplier of electricity for equipment connected directly to the generator by cords, plugs, or receptacles mounted on the generator expressly.
- The extraneous metal parts of the equipment (the parts not designed to conduct electricity) and the equipment grounding terminals of the receptacles are bonded to the frame of the generator.

The exceptions apply to both dedicated portable generators and vehicle-mounted generators. With regard to vehicle-mounted generators:

- The **frame of a motor vehicle** is considered sufficient grounding for vehicle-mounted generators incorporated into the vehicle's design and construction.
- OSHA regulations state that, in this instance, the generator must be **bonded** to the vehicle frame.
- OSHA regulations state that a vehicle-mounted generator grounded through bonding with the vehicle frame can only supply equipment located **on the vehicle**. Other means of permissible connection, in this usage, are through connection to mounted receptacles on the vehicle.

ONSITE ELECTRICITY GROUNDING REQUIREMENTS

The **Assured Equipment Grounding Conductor Program** is a written protocol which covers cord sets and receptacles which are not a permanent part of a building structure's electrical configuration. The safety program applies to cords, plugs, connectors, and electric-powered tools.

- The written program and the name of the designated "competent person" must be **available for review** by any affected employee or by OSHA officials.
- **Tests** are routinely required so as to assure workers that equipment and non-permanent electrical lines are adequately grounded. Required tests must be recorded and maintained until the next testing period.
- Two types of testing are required as part of the Assured Equipment Grounding Conductor Program:
    - A **continuity test** is needed to ensure that equipment grounding conductors are electrically continuous.
    - Another test must be done to ensure that the receptacles and plugs have the ground circuits connected to the **proper terminal**.

TEMPORARY POWER CORD SAFETY

Temporary power cords and extension cords should only be used on a short-term basis. They should not be installed as substitutes for permanent wiring. Any actual wiring work must be performed by a qualified electrical worker that has been trained under OSHA requirements and NFPA 70E. Extension cords should always be examined before use to ensure they are not frayed or the insulation missing and that the third ground prong is in place and not bent. Care must be taken to install the cords in such a way that they do not pose a trip hazard; for example, use a cord protector do not trip over it. If the cord will be used for electronic equipment or computers, it should have a surge protector in it to protect the electronic equipment against power surges.

TESTING TEMPORARY ELECTRICAL CIRCUITRY ON THE CONSTRUCTION SITE

Inspection of non-permanent electrical lines and equipment should be a routine part of the daily round on a construction site.

- All required tests must be performed **before** any work is performed when equipment is first used.
- Equipment should be tested again after it has been **repaired** and returned to the worksite. This is especially important when electrical parts are known to have been damaged.
- Routine inspections of electrical equipment must be re-tested within a **three-month period** unless cord sets and/or receptacles are part of a fix-wire system (in which case inspections must be within six months).

Test records must be maintained onsite and available for inspection. A record date of all inspections (and identification of the item being inspected) is required.

Electrical Continuity and Impedance Testing

On construction sites, all power tools and electrical lines which are not part of a building's permanent structure must be tested for **grounding and continuity of circuitry**. The types of equipment testing equipment which may be encountered for these purposes are:

- **Ground loop impedance tester**: This measures the amount of resistance or impedance in the ground terminal of power equipment used for drilling, sawing, lighting, etc.
- A **volt-ohm meter** would also measure the amount of impedance but the **impedance meter** provides data regarding both low and high ground faults.
- **Lamp and battery, battery and buzzer/bell, and neon lamp testing equipment** are used to measure continuity in circuitry but not the amount of resistance.

Arc Flash

Arc flash is an event in which electrical current jumps (or flows) through an air gap between two conductors; for example, broken or torn wire insulation can cause conditions that promote arc flash, as can dust buildup in an electrical panel. Arc flash is an extremely dangerous condition due to the tremendous energy released during the event. The arc flash explosion can cause severe burns, and the explosive concussion can throw an employee some distance from the event. It can also cause hearing damage. Unfortunately, the susceptibility of a given electrical panel to arc flash is not apparent to the naked eye, making an arc flash assessment and electrical panel labeling program important.

All electrical workers must be trained in the basic elements of arc flash safety. All electrical panels must be evaluated by qualified personnel to determine the arc flash potential. This evaluation considers the potential release of energy that would occur (based on voltage and load) were there to be an arc flash in that panel. From this, the panel is rated on a scale of 0 to 4. Personnel that will be working in the panel must wear personal protection suits rated for the appropriate arc flash hazard level. In addition, workers must be trained to set up appropriate exclusion zones around areas where work will occur on electrical panels. Workers must be trained in emergency procedures and first aid procedures for victims of electrical shock and burns.

Common Hazards and Controls Associated with Excavations

Underground Excavations

A competent and knowledgeable person must assure that workers receive proper training when working in **underground areas**. The requisite areas of required training demand that all underground workers be taught to recognize and avoid hazards which are peculiar to underground construction. Those areas where training must be applied are:

- Employees must be taught the methods of **air monitoring** and to recognize potential indications of contaminants.
- Workers should know the air **ventilation standards** which apply and the required amounts of **illumination**.
- The use and methods of **communication equipment** must be taught to workers. Effective means of communication from excavation to the above-ground area must be provided.
- Employers must be trained in **flood control, fire control, and fire prevention**.
- **Personal protective equipment** training is of vital importance. Familiarity with emergency breathing apparatuses is part of this training.
- Workers must be fully informed regarding **check-in and check-out methods** as well as **evacuation plans**.

The employer is required to take swift action when a competent person has determined the presence of hazardous gases in an underground excavation:

- **Notices** of the gaseous condition must be posted at all entrances to the excavation.
- If the hazardous gases amount to at least **five percent** of the LEL, **ventilation air volume** must be immediately increased to reduce the concentration to an amount less than five percent of the LEL. There is an exception when operating under gassy/potentially gassy requirements.
- Work must be entirely **stopped** when welding, cutting or other hot work is being done if the concentration is **ten percent** or more of the LEL. Concentrations must be reduced by ventilation until the concentration is below ten percent of LEL.

Underground tunnels, shafts, or underground cavities

Underground work has a wide array of dangerous conditions which must be addressed in worker training. While some of the areas of instruction may be specialized, as in the use of electronic testing equipment, all workers should be informed of several general hazards:

- **Illumination**: Workers should be informed of appropriate light levels when working underground.
- **Communications**: A clear understanding of the communications systems to be used in both routine work and/or emergency conditions is required.
- **PPE**: The use of personal protective equipment (SCBA) is necessary in many cases. Workers must be trained in the potential hazards requiring respiratory protections.
- **Fire Prevention and Protection**: Whether the activity is routine or specialized, such as in the use of explosives, workers should be trained sufficiently in these areas so as to contribute to overall safety of all workers.
- **Air monitoring**: Air quality cannot always be detected without instruments but workers should be trained to recognize certain indications of developing hazardous conditions.
- **Ventilation**: Workers should be educated to understand individual requirements and the methods used to maintain proper ventilation.
- **Floor Control**: Water seepage into shafts, tunnels and excavations is a common risk factor. Workers must be trained in emergency evacuation procedures.

Regulations designed to prevent injury when workers are required to work underground

The OSHA provisions in **1926.800** provide the basis for worksite safety requirements of underground workers employed in the excavation of underground cavities, tunnels, shafts or chambers. The general provisions of those requirements are:

- Underground areas must be properly **vented** to prevent accumulation of dusts, fumes, vapors, and gases.
- The number of employees working underground should determine the minimum capacity of ventilation requirements. **Two hundred (200) cubic feet per minute of fresh air** is required for each employee.
- **Mechanical ventilation** is a must in most underground circumstances. The velocity of air must be regulated to flow at least 30 feet per minute in underground areas where conditions are likely to create dust, fumes, or gas vapors.
- **Air flow direction** should be reversible.
- If **blasting** is to be conducted in an underground tunnel or cavity, workers cannot return to work until the ventilation system has cleared the area of all exhaust and fumes.

- **Ventilation** doors must be designed to remain closed when in use regardless of the direction of air flow. Ventilation doors are designed to provide a barrier to dangerous fumes should they build up in adjacent areas.
- If the ventilation system has been shut down for any purpose, **testing** must be done by a competent person to determine whether the area is safe for reoccupation by workers.
- **Entry/exit openings** must be strictly controlled and posted with appropriate safety signs.
- Whenever any employee is working underground, a "**designated person**" must be present above ground. The "designated person" is required to keep a running account of the number of underground workers and to secure and communicate a rescue response when necessary.
- It is the responsibility of the employer to **communicate and coordinate** with other workers and/or job operations which may impact the safety of underground workers. Movement of heavy equipment is an example of other worksite activity which may impact underground workers.
- OSHA section 1926.800 mandates that **rescue plans** must be established in the event of emergency. Specific requirements are based on the number of workers employed underground.

OSHA section **1926.800** requires contingency plans for the emergency use of rescue personnel on construction job sites.

- Whenever employees working underground number 25 or more, regulations require that the employer make **contingency plans** with locally available trained rescue personnel.
- When 25 or more employees are working underground, contingency must be made for **two rescue teams**. One team must be located within a half-hour traveling distance; the other must be within a two-hour travel time period.
- If less than 25 persons are working underground, contingency plans may involve a **single rescue crew** which must be positioned within one-half hour travel time from the job site.

Proper contingency planning requires that rescue personnel be made familiar in advance of conditions at the job site.

CONSTRUCTION TRENCHING

When a **trenching system** is to be dug to a depth of twenty feet or more, a professional engineer is required to design a safe, strong, and reliable method of **shoring up the excavation**. The CHST must be familiar with ground materials and slope requirements for trenches less than 20 feet deep. Soil types are categorized as Types A, B, or C. The maximum angle of a trench wall depends on the category type into which the excavation has been made. Clay is more stable than other ground materials like sand or silt and is considered a Type A material. The maximum slope of an excavation into Type A soil is 53 degrees. Gravel and silt are Type B materials. The maximum height to depth ratio for trenching in Type B soil is 1:1, yielding a maximum 45-degree angle. Type C is sand, which is very unstable for trenching. The slope angle for a trench into Type C soil can be no more than 34 degrees. The only material in which vertical trench walls are allowed is solid rock.

If sloping is undesirable, **shielding** may be used instead. Various types of shield systems may be designed to protect workers. These may range from portable systems which may be moved from one site to another or systems manufactured on the spot and according to OSHA specifications. "Trench boxes" or "trench shields" are terms which may be used to describe shielding used to protect workers in trenches.

Physical Conditions Which Apply When a Soil or Earth is Reported as Distressed

A "distressed" soil or area of earth and rock is one that poses an especial danger to workers in the vicinity. There is an imminent and greater potential for a collapse and injury to workers occupied in an excavation of "distressed" soil. Distressed soils can be recognized by:

- **Fissures or cracks** are apparent in the face of or on the sides of an open excavation.
- When the edges of an excavation show signs of **subsidence or sinking**.
- When the bottom of an excavation shows signs of **bulging or separating**.
- **Spalling** means that a material is breaking up into small chips, flakes, or splinters. It is an indicator of distressed earthen materials.
- Pebbles and little clumps of material **separate** from the face of the side cut and roll down into the excavation.

Conducting the Required Daily Investigation of an Ongoing Excavation Site

Daily inspections of excavation sites are required as a minimum but inspections can be made at any time a change in work, weather, or environmental conditions makes more frequent inspections desirable. The safety inspector must be alert to all hazardous conditions affecting worker safety but there are certain commonalities of concern:

- Examine for failure of protective systems, hazardous atmospheres, water accumulation, and the ground signs which may indicate a potential collapse of excavation walls.
- Means of **exit or egress** should be inspected. Stairways, ladders, ramps or other safe means of evacuation from trenches are required in ditches or trenches that are 4 feet or more in depth. A worker should not have to travel more than 25 feet to reach a means of safe exit.
- **Stability** of adjoining walls, buildings, or other nearby structures should be inspected for settling, leaning, or other signs of tipping.

Above-Ground Utilities Likely to be Affected by Construction Site Excavation

Above ground utilities are usually visible and are easily mapped in preparation for site excavations. Though they are clearly visible in most instances, above ground installations of utilities require caution and preparation on the part of excavation planners:

- **Trucks and other vehicle drivers** passing through the site must be adequately warned of the hazards of above-ground utility lines. **Operators** must be briefed beforehand. Caution signs at planned traffic locations increase safety significantly. Drivers should be forewarned as to the height of their vehicles or excavating equipment.
- In the planning stage, great care should be given to the **route of traffic** which must bypass the planned excavation. Traffic from heavy equipment must not be allowed to pass in such proximity to the excavation that might cause undue pressures upon the walls of the excavation and trigger cave-ins.

Underground utilities likely to be affected by construction site excavation

Preparation for site excavation in areas where underground sewer, power, phone, or gas lines may be buried requires:

- Individual **utility companies** must be contacted and apprised of the nature and specific location of the proposed work. Requests should be made for information pertaining to the specific location of underground installations.
- If the individual utility does not respond to requests within a **24-hour period**, the construction manager may begin work on the excavation, exercising all caution. The exercise of caution means that detection equipment to locate lines is routinely used, along with other acceptable measures.
- **Mapping**: The estimated location of all underground utilities should become part of the site map drawn up prior to beginning the excavation. Sufficient numbers of copies should be distributed at briefings to managers, equipment operators and laborers.

Training requirements

As part of a comprehensive injury prevention plan and hazard communication plan, workers who work in an excavation site must be trained on the hazards of an excavation and how they are protected from injury. They should be trained in site-specific emergency procedures. If wearing a harness, the employee must be trained in how to use the harness and how to inspect it to ensure it is working optimally. If the excavation meets the definition of a confined space, the employee must have confined space training.

Safety requirements

The risk of a cave-in is the greatest risk at an excavation site. Any excavation of greater than five feet deep must have a protective system in place to prevent cave-ins (barricades, sloping side, benching, etc.). If the excavation is greater than twenty feet deep, this protective system must be designed by a registered professional engineer. There are several types of cave-in protection available, such as shoring the sides, sloping the sides to reduce the vertical edge, benching the sides, and shielding workers from the sides. A qualified individual should be consulted when choosing cave-in protection, as site conditions such as soil type, moisture level, and the amount of activity around the excavation must be taken into consideration. Access and egress to the excavation area and the excavation itself must be controlled with barricades and work practice controls. A ladder or ramp must be provided as a means of egress if the depth of the excavation is four feet or greater. The route to the means of egress cannot be more than 25 feet for any employee who may be working in the excavation. The spoil pile (the soil removed from the pit) must be set back from the edge of the excavation at least two feet. This prevents material falling on workers and makes an even walking surface at the edge of the excavation.

COMMON HAZARDS AND CONTROLS ASSOCIATED WITH WORKING AT HEIGHTS

SAFETY NETS

According to the provisions of OSHA 1926.502, **safety net systems** should be constructed with the following considerations:

- Safety nets should not be positioned more than **30 feet below the working level**. Nets should be placed as close to the work area as the activity permits as long as they are unobstructed.
- Safety nets should be positioned so as to extend **beyond the vertical plane** of the working surface at a distance of from 8 to 13 feet, depending on the distance of the net from the working area.
- Safety nets should have sufficient **"ground" clearance** so as not to strike surfaces below the working level. This precaution is determined by **drop testing** and is the responsibility of the employer. Drop testing is a significant factor in determining the strength of the net and its positioning.
- Drop testing consists of a **400-pound sandbag** dropped from the highest working surface to which workers are exposed.
- **Scrap materials and/or tools** which have fallen into the safety net must be removed expeditiously but certainly no later than the start of the next work shift.

SCAFFOLDING

There are many different types of scaffolding designed around specific construction uses. The terminology used to describe scaffolding is varied. Therefore, it behooves the CHST to know both slang terms for the various types as well as the technically correct names provided by the manufacturer.

- **Horse Scaffold**: This is merely a platform supported on each end by construction "horses" or "saw horses." When made of metal, these are sometimes called trestle scaffolds.
- **Chimney Hoist**: The technical name for a "chimney hoist" is multi-point adjustable suspension scaffold. This type of scaffold is used to provide access to the inside of large chimneys.

The **fabricated frame scaffold** is a common type made of tubular steel and welded at the assembly joints. Aside from the mainframe, the fabricated frame scaffold consists of other components like horizontal bearers and intermediate bearing braces or members. **Bricklayers' square scaffolding** is constructed from a system of framed squares which provide a solid, wide, and stable footing for masons and tenders. A **carpenters' bracket scaffold** attaches to the building or the stable vertical walls of a building under construction. A **catenary scaffold** is a standing platform suspended from parallel horizontal ropes or cables attached to beams or other solidly supported structures. Very often, the catenary scaffold is supported by additional vertical hoist and pulley system. A **hoist** is a manually operated or power-driven device which attaches to a **suspension scaffold**. It is the mechanism which allows the suspended platform to be pulled up or "hoisted" aloft. There are two common safety devices which are designed to work in conjunction with a hoist to augment worker safety:

- **Lifeline**: Scaffolds and suspension devices must be anchored to solid structures to prevent swaying, settling, or otherwise collapsing. The lifeline is a rope, chain, or cable that ties a platform to strong and stable anchor points. A lifeline may be positioned vertically or horizontally and fastened at several points.
- **Decelerating Devices** are mechanisms specifically designed to augment scaffold safety systems. Their chief purpose is to dissipate energy which would harm a falling worker attached by harness or belt. A decelerating device protects against the shock or whip action imparted to a falling body as it strikes the end of a safety line or other implement. It slows and gentles the falling motion.

The terms refer to safety devices or components associated with scaffolding or suspension mechanisms which allow workers to conduct safer activities alongside or upon the heights of buildings.

- **Stall Load**: The "stall load" refers to the point at which a power-operated hoist is automatically disconnected for worker safety. Stall load is designed into the mechanism to prevent overload. Should a worker overload a suspension scaffold, the operating mechanism disconnects until a proper load weight is restored and amendments are made.
- **Rated Load**: Rated Load is the maximum load for which a hoist or scaffold is designed to accommodate. Rated load of a scaffolding mechanism can be obtained by reading the manufacturer specifications.

REVIEWING AND INSPECTING SAFE UPRIGHT STANDING SCAFFOLDING PROCEDURES

Since scaffolding is commonly used and because there are so many types of scaffolding, the CHST would have a long checklist of safety items to review when assessing **worker safety**:

- All scaffolds must be **assembled** by a competent person capable of supervising the placement, disassembly, and moving of scaffolding.
- The competent person must have knowledge of **guardrail specifications** for tubular welded frame scaffolds and for the manner in which they are to be braced.
- Mobile scaffolds must be tightly planked with **scaffold grade planking**. Planking which extends beyond scaffold supports must meet additional OSHA specifications.
- The base of any scaffolding must be set upon a **sound and stable base**. The positioning of upright supports must be plumb and the joints securely locked.
- Safe access by means of a **secure ladder or permanent stairway** must be provided. Ladders must be affixed to the scaffold frame. The scaffold frame is to be anchored into solid and stable side structures at points along its heights.

CIRCUMSTANCES UNDER WHICH LADDERS CAN BE USED UPON SCAFFOLDING PLATFORMS

As a general rule of thumb, ladders should not be used to increase the working height of employees on elevated scaffolding. There are exceptions, however:

- Where there is a large area scaffold designed to be used as a temporary "floor" and wide enough to provide security, a ladder may be used to elevate the working height.
- When ladders are used upon scaffolding, the structure is subject to a sideways thrust. Under these circumstances, the main body of the scaffolding must be firmly anchored to walls or secure beams.

- The ladder's legs must be solidly planted and blocked to prevent skidding or slipping.
- All other moving parts of the scaffold must be secured against movement or shifting.

Measurement rules for setting ladders against the walls of a work area

There are three elements to be considered when observing the regulations regarding setting a straight ladder against the walls of a work area.

- The **base** of the ladder refers to the distance between the wall and the bottom of the ladder. This distance should be no more nor less than 1/4th the length of the ladder from floor to top point.
- The **length** of the ladder in this application does not mean the entire length of the ladder but refers to the distance from the point where it touches the floor to the point where it is supported by the wall.
- Ladders should be firmly **anchored** with stops in place and secured with blocking.

Load ratios for scaffolding and suspension lines or ropes

Scaffolding load ratios require that scaffolds be constructed with the capability of supporting many times their intended load.

- **Scaffolding and planking** must be able to support at least four times the maximum intended load.
- **Suspension ropes** must be capable of supporting at least six times the maximum intended load.

There is good reason for the OSHA requirement that scaffolds and ropes be able to support far more than their intended load:

- In addition to the workers using the scaffold, there is the added weight of **tools and equipment** to consider.
- The effect of the **elements** like snow and wind can have an impact on the carrying capacity of scaffolding.
- **Planking** may be weaker than the specifications would indicate, particularly if the lumber is of lower quality and contains knots or fissures.

Personal fall arrest system

A personal fall arrest system is a method of stopping an employee from falling from the heights at which he is working. Some personal fall arrest systems involve harnesses, hooks, belts, and loops of rope or other suspension mechanisms designed to catch a falling worker. Fall arrest systems must:

- Not subject an employee to a force greater than 900 pounds if a **body belt** is used.
- Not subject an employee to a force greater than 1800 pounds if a **body harness** is used.
- Not allow an employee to fall more than **six feet** nor come in contact with the **floors or objects below**.

Personal fall arrest systems must also be able to sustain two times the impact caused by an employee falling six feet. The deceleration distance of travel can be no longer than 3.5 feet.

Positioning Device Systems

Several types of construction activity require the use of positioning devices which allow the worker to hang suspended by harness or belt upon the area being worked on. OSHA regulations require that:

- The **rigging** for the positioning device be designed such that an employee cannot fall more than two feet.
- The point to which the positioning device is **anchored** must be stable and capable of supporting the impact of a 3000-pound inertial force.
- Connectors should be of **drop forged** or formed steel or similarly strong materials. They should be corrosion resistant and smooth with a tensile strength of at least 5000 pounds.
- Dee-rings and snap hooks must have a minimum **tensile strength** of 3600 pounds.

Positioning device systems must be inspected for wear prior to use. Equipment showing signs of wear or damage must be replaced immediately.

Height at Which Fall Protection Systems are Required

Personal fall arrest systems (harnesses, for example), guardrails, and safety net systems must be used whenever workers are employed in construction or walking in work areas six feet or more from the lower level.

OSHA regulations require that the "six foot" rule be applied to the following situations:

- Workers must be protected from falling through **holes or floor openings** like skylights or stairwells. Skylights may be covered with solid structural materials. Another method may be to erect a guardrail system around floor openings or skylights.
- Workers employed in **framework or reinforcing steel frame construction** must be protected by personal arrest systems and/or safety nets.
- Workers who must walk through or work near **trenching areas or excavations** must be protected by fences, barricades, or guardrail systems if the excavation is six feet or more.

Providing Greater Fall Protection to Construction Site Workers

Common errors are often repeated around construction sites so it is important to recognize the following **fall protection strategies**:

- The assignment of a guard for **floor holes** is necessary especially when an area is being used as a walkway by materials handlers who may not be able to see over the materials they are carrying.
- Body belts, lanyards, and fall-breaking mechanisms must be employed while working from **aerial lifts**.
- **Wall openings** must either be guarded or securely barricaded. The same strategy may be employed in preventing falls from open-sided floors or platforms.
- When no other practical means of fall protections is practical, **safety nets** must be securely arranged to protect workers at heights of 25 feet or more.

SAFE PRACTICES FOR USING FALL PROTECTION SYSTEMS

For effective fall protection, OSHA recommends that companies adhere to the following practices:

- The company should have a written fall protection plan as part of its overall health and safety plan. The plan should include company rules for how and when to use fall protection equipment.
- The company should follow standard fall protection requirements when fall protection equipment must be used, usually when an employee in a general industry is four feet above the floor, when an employee of a construction company is six feet above the ground, or when an employee is on scaffolding 10 feet above the ground.
- The company should provide correct fall protection equipment and ensure that it is not only used, but is used properly.
- The company should inspect, maintain, repair, and replace fall protection equipment regularly.
- The company should provide supervisors and workers with training on how to recognize fall-related hazards and how and when to use fall protection equipment.

A fall protection system can limit or prevent falls. A fall protection system can include safety belts, safety harnesses, lanyards, hardware, grabbing devices, lifelines, fall arrestors, climbing safety systems, and safety nets. Most of these elements stop a fall that has already started and must meet specific standards. Safety belts are worn around the waist while harnesses fit around the chest and shoulders and occasionally the upper legs. Safety harnesses lessen the number and severity of injuries when they arrest a fall because the force is distributed over a larger part of the body. Lanyards and lifelines connect safety harnesses to an anchoring point while grabbing devices connect lanyards to a lifeline. Lanyards absorb energy, so they reduce the impact load on a person when the fall is arrested.

SAFE WORK PRACTICES WHEN WORKING AT HEIGHTS OR ON ELEVATED WORK PLATFORMS

Elevated work platforms or working at heights pose a fall hazard that can be fatal if an employee falls on his or her head. When using a platform, there should be guardrails that conform to Occupational Safety and Health Administration (OSHA) guardrail guidelines. Employees should be trained in and use fall protection harnesses; the harness must be tied to a place that will safely arrest the fall and not tip over any equipment. If the height is accessible using a ladder, make sure the ladder is on a stable surface and that a coworker is available to hold the ladder and position it so that reaching from the ladder is not necessary. Never stand on the top step of the ladder.

AERIAL LIFTS AND SCISSOR LIFTS

An aerial lift is a basket on an extended arm for lifting workers to heights. A scissor lift accomplishes the same thing but is conveyed upward vertically. Both have guard rails around the platform. Some safety professionals believe that a worker using an aerial lift or a scissor lift should also wear a safety harness in case of falls. However, this is not a requirement according to an OSHA interpretation issued in a letter dated July 21, 1998, as long as the lift is equipped with an OSHA-approved guard rail. In fact, using a safety harness to arrest a fall from a lift can be an additional hazard if the force of the fall causes the apparatus to tip over onto an employee.

COMMON HAZARDS AND CONTROLS ASSOCIATED WITH WORKING IN CONFINED SPACES

PERMIT-REQUIRED AND NON-PERMIT-REQUIRED CONFINED SPACES

There is a significant difference between a permit-required and non-permit required confined space. Confined spaces are areas not designed for continuous employee occupancy, and which have limited means of egress. The definition includes spaces that even arms are placed into, not just spaces that can accommodate the entire body. They may also have the potential for a hazardous atmosphere, either oxygen deficiency, presence of chemical vapors, or extreme temperatures. For this reason, Occupational Safety and Health Administration (OSHA) requires assessments and inventories of potential confined spaces to determine appropriate entry procedures. Entry into a confined space using a permit system is required when there is a potential for oxygen deficiency, explosive atmosphere, and/or chemical vapor exposures. The permit provides a mechanism to track entries and to document that the proper pre-entry procedures have been followed, such as measuring oxygen levels and measuring chemical vapors. Employees must also have an attendant outside the confined space at all times and emergency equipment on hand in case of emergency.

Permits issued for this purpose must:

- Identify the space to be entered.
- Identify the type and nature of the work to be performed and provide substantial justification for issuing a permit.
- Be date stamped with the date of authorization and the number of hours or days in which confined-space work has been authorized.
- Identify the person or persons who are authorized to enter the confined space. This section of the permit should specify the type of tracking and monitoring system used to control entrance and egress.

WORKING IN PERMIT-REQUIRED CONFINED SPACES

A confined space has the following characteristics:

- An area sufficient in size as to accommodate a worker of normal size to enter and perform assigned work with necessary tools and uniform.
- A confined space has limited entry and/or egress points which are so constructed as to limit the free passage of employees or rescuers into the area. Holding tanks, grain silos, manure pits, and entry passages to sewage tunnels are confined spaces requiring special methods of access and preparatory measures taken to ensure health and safety under poor atmospheric conditions.
- A confined space is one in which continuous employee occupancy for work purposes cannot be expected to last for an entire 8-hour period of time as would be the case under normal working conditions. OSHA limits the amount of time a worker may be exposed to confined spaces with limited or extremely hazardous atmospheres.

CONFINED SPACE PROGRAM

A confined space program must contain an inventory of all confined spaces at a site, a formal documented assessment of the potential hazards posed by the space (e.g., Is there a danger of oxygen deficiency? Are there potentially hazardous vapors or fumes present? Is there a risk of high or low temperature environment?). A confined space entry permit must be developed and required to be used for all permit-required confined spaces. All confined spaces must be labeled as such, and the label should identify whether the space is permit required or not. The employees that will enter the confined space must receive appropriate training in the entry procedures and what appropriate emergency procedures are.

To fully implement a confined space safety program, one must be able to evaluate whether a hazardous atmosphere exists inside the confined space. The first assessment that must be made is to determine whether there is an oxygen-deficient atmosphere. For this, an oxygen level meter is needed to confirm that the oxygen content of the air inside the confined space is at least 19.5 percent. Second, if there is any possibility that flammable dust or vapors exist, the air must be analyzed for either dust or organic vapors. Organic vapors can be analyzed using a handheld organic vapor analyzer, and the results can be used to establish whether respiratory protection is necessary. The confined space entry plan and testing plan should ideally be overseen by an industrial hygienist to ensure that no potential hazards are overlooked.

Controls for Working in Confined Spaces

Confined spaces include such work areas as tank cars, boilers, silos, underground tunnels, and railroad boxcars. All these spaces have limited entrances and exits and require specific controls to ensure worker safety. Hazards that workers in confined spaces face include toxicity, potential oxygen deficiency, and fire or explosion from flammable or combustible gases or dust. To protect workers, the following actions should be taken:

- Always evaluate a confined space for hazards before workers enter.
- Ensure that the confined space has adequate ventilation.
- Include equipment for suppressing fires and removing smoke and fumes.
- Train workers on safety procedures they need to follow when working in a confined space.
- Institute a buddy system for confined spaces so two workers are always present.

Common Struck by Hazards and Associated Controls

OSHA defines 'Struck by' as an injury caused by the contact or impact of a piece of equipment or object and the injured person. The impact of the object with the person is the cause of the injury. Types of 'Struck by' hazards include: flying objects, falling objects, swinging objects, and rolling objects. These can cause severe injury. Control measures to prevent struck by injuries include enforcing height limits on pallet stacking, toe boards and other barriers, proper training for those operating various equipment and vehicles, etc. Finally, proper PPE is critical, especially hard hats. The hard hat is the best piece of PPE to prevent 'Struck by' injuries due to falling objects.

Flying Objects

Flying objects are a common occurrence at a construction site. Activities that may cause flying objects include **demolition activities** such as wall removals or removal of other infrastructure. Flying objects can also occur if **compressed air** is used to clean up the work site; for example, nails, wood, and concrete debris can become airborne if compressed air is used. Best practices for preventing injuries from flying objects include mandating that all workers wear **eye protection** (safety glasses or goggles) on the work site, along with a **hard hat**. High-risk activities should require the worker to wear a **face shield**. Compressed air should not be used as a method of clean up due to the risk of flying objects and dust. Instead, use vacuums and sweeping.

Falling Objects

Overhead work is common at a construction site. Overhead work could include work on ceilings or roofs, or having materials transported overhead by cranes or hoists. Measures that can be taken to reduce the risk of injury from being struck by a falling object are to **secure** any tools and raw materials by means of fastening them to a fixed object (such as scaffolding) to prevent them from falling. **Debris nets** can also be used beneath the work area to catch anything that might fall. Additional measures include providing **barricades** (safety cones, caution tape, warning signs) to restrict entry beneath and around the work zone. **Personal protective equipment** (hard hats) should always be worn on a construction site.

Swinging Objects

Swinging objects can be found on a construction site when transporting materials overhead using a **crane** or **hoist**. For example, beams or joists must be transported overhead during installation. Best practices are to first **plan the overhead move** and ensure there are no other activities occurring in the immediate vicinity at the same time. All job site participants need to be aware of the overhead activity and the potential for a swinging object. The immediate area under and around the overhead activity should be barricaded off to restrict entry beneath the crane or hoist move. Ensure that all inspections are done on the crane or hoist prior to doing the overhead lift. Also ensure that the movements are done in a slow and controlled fashion to prevent excessive movement of the load.

Rolling Objects

Rolling objects at a construction site can include items such as roofing materials that are supplied by the roll, or a vehicle that escapes control. Controls for being struck by rolling objects include being **aware** of the possibility and ensuring that movements of these items are **planned and controlled**, and discussing the potential for being struck by these objects at daily safety tailgate talks. Ensure that large rolls of material are secured before moving them, and ensure that there are no workers in the movement path. Workers should not walk in front of moving vehicles but should follow behind. Vehicles should be parked, when unattended, with parking brake engaged and should not be parked on a slope.

PPE for Struck-By Hazards

The best practice at a construction site is for all employees and visitors to wear **hard hats** that are approved by the **American National Standards Institute (ANSI)**. OSHA requires hard hats to be worn whenever there is a risk of being struck by falling objects. Two types of hard hats are available: Type I and Type II. **Type I hard hats** are designed to protect from the full force of the impact directly on the top of the head, whereas **Type II** will also protect from a blow that is received on the side of the head or off-center, in addition to one received on the top. Be aware that the suspension of the hard hat should be replaced annually to ensure top performance.

Planning for Overhead Lifts

Using a crane or hoist to perform an overhead lift requires planning. Many factors must be taken into account. For example, what other activities are occurring at the same time? Is any accommodation or rescheduling needed to ensure that the lift doesn't cause additional hazard? Roles and responsibilities for the lift need to be planned, as well as how to communicate during the lift. The capacity of the crane or hoist to carry the weight of the load needs to be verified. The stability of the crane or hoist needs to be ensured. The path the crane or hoist will take needs to be planned out to ensure there are no structures in the path that will be impacted. The weather needs to be considered; if there are high winds, it needs to be determined whether this will cause unintentional swinging of the load.

Common Caught-In or Caught-Between Hazards and Associated Controls

OSHA defines 'caught-in or between' injuries as harm caused by a person being squeezed, caught, crushed, pinched, or compressed between machinery or parts of equipment. This type of hazard and injury can cause broken bones, amputations, strangulation, or internal injuries. The best control measure for this type of risk is to construct physical machine guards that prevent any body parts from insertion in or near the equipment. Also, workers can help mitigate potential harm by not wearing jewelry (necklaces, and other things that dangle) or loose clothing (un-tucked shirts, etc.).

Control Measures for Heavy Equipment

Heavy equipment is common at a construction site. To avoid being crushed by heavy equipment, it is important that workers never position themselves between a wall and heavy equipment. There must also be designated and barricaded **work areas** for heavy equipment where pedestrian traffic is not allowed, and heavy equipment must be equipped with **back-up audible warning equipment**. Recognize the potential for heavy equipment to tip over and crush the driver. Control measures include observing the equipment's load carrying limit, wearing safety belts when operating the equipment, and verifying and observing the degree of slope the equipment can be operated on.

Cave-in Protection

Trenches can be dangerous due to the possibility of cave-ins. If the trench is **greater than five feet deep**, it is required to be shored up by an engineered protective system unless it is constructed solely of stable rock. Trenches can also be protected from cave-in by sloping the sides to reduce risk of the walls collapsing. Devices called **trench boxes** can be used to shore up the sides and stabilize them. Any shoring or sloping of a trench must be overseen by a competent and trained person. Trenches must be inspected daily for signs that the walls are stable, and all workers should be informed of trench safety practices. Heavy equipment should not be operated near the open edge of the trench due to the possibility of weakening the walls and causing a cave-in.

Control and Inspection Measures for Scaffolding

Scaffolding is commonly used in the construction industry to access elevated parts of buildings and to provide a work platform for workers and tools. Scaffolds must be constructed to hold at least **four times the expected weight** on the platform. Boards must be of **proper grade** (for example, no plywood) and must not overlap with one another; that is, each board must rest on its own support rail. **Guardrails** must be in place, and **fall protection** (harness) must be used if the scaffold is more than ten feet high. The scaffolding should be inspected daily to ensure that the construction is still sound and the boards do not show signs of wear, and that there is no debris or ice on the scaffolding.

Machine Guarding

Machine guarding is the practice of placing **physical barriers** on tools or machinery to prevent human contact with moving parts that can damage body parts. An important part of a **tool inspection program** is examining machine guards for their presence and ensuring they are not removed. On construction sites, commonly encountered machine guards include **blade protectors** on circular saws and **guards** on grinding wheels. Guards must also be in place when there is a risk of exposure to rotating shafts, belts, or gears.

Common Hazards and Controls Associated with Hoisting and Rigging

Personnel Hoisting

When conventional means of moving workers to higher levels are not feasible, a crane can be used to hoist personnel, so long as the following cautionary procedures are observed:

- The crane must be on level ground. The angle between crane base and the ground cannot exceed **one percent**.
- Total weight of personnel hoisted plus the weight of the platform itself cannot exceed **fifty percent of the rated capacity** of the crane or derrick.
- If workers must be hoisted aloft by crane, the **platform** should be moved in a slow, cautious and controlled manner.

- **Carrying load lines** must be capable of supporting seven times the weight of the load. If anti-rotation rope is used, it should be of a type capable of supporting ten times the intended carry weight.
- All **brakes and locking devices** on the crane must be engaged when the personnel platform is stopped.

PERSONNEL PLATFORMS

When conventional means of hoisting personnel to higher building levels are unavailable or infeasible, a **personnel platform** may be used. OSHA regulations require that such platforms be designed by a qualified engineer or other qualified competent person.

- The platform must be designed to **minimize tipping** when workers move or shift their weight on the platform.
- The platform itself (without guardrails or fall arrest systems attached) must be able to support **five times the maximum intended load**.
- **Guardrails and grab rails** should be installed around the entire perimeter of the platform.
- **Gates** for access must not be outward swinging. A fail-safe device must be installed to prevent accidental opening of the gate or entrance.
- Platform or cage must be of a height permitting employees to stand **upright**.
- The platform or cage must be provided with **overhead protection** to prevent injury from falling objects.

Personnel platforms are designed by an engineer or other qualified person to meet specific OSHA requirements. The load rating must be specified on a plate attached to the platform itself. However, these precautions must be augmented by an attempted lift or "trial lift" at the rated load capacity.

- The platform must be loaded to the **weighted load capacity** (without people) and lifted from the ground.
- During this trial lift, the hoist operator must **test** all controls, systems, and safety devices to see if they are properly functioning.
- A single **load test or trial lift** is sufficient so long as all systems are properly functioning and the crane or hoist is not moved.
- If the crane or hoist must be moved and used in a different area, a **new trial lift** must be conducted at the new location.

After the trial lift of a personnel hoisting platform, a competent person must immediately conduct a **visual inspection** of the crane or hoist to see if any defects resulted from the lift. Among the areas for inspection are:

- The **platform** must be examined to see if it is properly balanced and does not tilt to one side.
- **Hoist ropes** must not be twisted, frayed, or entangled.
- The **linking mechanism** between the hoisting lines and the platform itself should be centrally positioned so as to prevent tilting.
- **Hoisting lines** should be inspected for slackness. Slack lines may indicate that hoist ropes or cables are caught or hung up in drums.

A good safety and health monitoring program at a construction worksite requires constant attention. There are many events which might cause personnel hoist operators to lose focus. Some of the **unsafe conditions** a CHST might observe are:

- A crane operator leaves the cab of his equipment to talk to a co-worker while workers are suspended in the platform.
- A harried site manager urges the crane operator to continue hoisting personnel aloft even when a dangerous wind has begun to blow.
- The crane operator loses communication and visual contact with the workers in the hoist.
- A group of workers takes a shortcut beneath the crane and platform boom in order to reach the area where they've parked their cars.

Safety in the use of personnel hoisting systems is a combination of **preparatory precautions** (proper construction of the platform) and **operating precautions**. With regard to operating infractions, the CHST might notice the following unsafe practices during crane or hoist operation:

- An employee reaches his arms outside the perimeter of the platform or cage during raising and lowering. Employees must remain entirely inside of platforms during operations.
- An employee tries to open the gate or access door to the platform before the platform is landed. The platform must be properly landed and secured to the structure before exit operations.
- Tag lines are not being used.

Pre-job Inspection of Rigging Equipment

A pre-job safety inspection must be conducted before any rigging job. Ensure all lines to be used are not frayed or broken. Ensure all employees are aware of how the job is planned to be conducted and are aware of their respective roles. Ensure the controls on the equipment operate as planned and that the equipment is parked in a stable surface and will not rock once the load is engaged and lifted. Inspect hydraulic lines for leaks, and ensure there are proper levels of hydraulic fluid. Evaluate the need for tag lines, and employ them if necessary.

Tag Lines

Tag lines are used in the use of cranes and hoisting equipment to attach to and stabilize the object being lifted. Lifting and moving an object with a crane will cause the load to swing on the end of the cable like a pendulum, which is potentially dangerous to employees. The tag lines are affixed to the load and to the ground to stabilize the load and prevent it from swinging. The tag lines must be made of soft material such as nylon or sisal, not of wire rope. Tag lines are required when the risk to employees of being hit by a swinging load exists or when there are overhead electrical lines that may be hit by a swinging load or crane.

Wire Rope Used in Construction Rigging

The use of wire rope for construction rigging must meet OSHA specifications:

- There cannot be more than **three full tucks** in a wire rope. Other permitted types of connection may be used if they meet safety specifications.
- Wire ropes used for hoisting or lowering must be constructed of a **continuous length of rope** without knot or splice except for the eye splices at the end of the rope.

- The eyes in wire rope slings or bridles must not be comprised of **wire rope clips or knots**.
- Wire rope shouldn't be used if the total number of **visible broken wires** is in excess of ten percent of the total number of wires. Nor can wire rope be used if the rope shows other signs of excessive corrosion or wear.

Hand Signals for Communicating in a Rigging and Hoisting Job

Hand signals should be used whenever the crane operator cannot see the load or landing area or cannot see the path of travel or when the load is close to power lines. Signals should be communicated bare-handed. The common signals are as follows:

- BOOM UP—Raise the arm at the elbow with the index finger pointing up, then move hand in a horizontal circle.
- BOOM DOWN—Lower the arm straight down with the index finger down, then move the hand in a horizontal circle, pointing down.
- TRAVEL FORWARD—With palm up, move the thumb in the direction of travel.
- STOP—Extend the arm straight out with palm down.
- EMERGENCY STOP—With the arm extended straight out, palm down, move rapidly right and left.

Hazards and Controls for Hoisting Apparatus and Ropes, Chains, and Slings

Hoisting apparatus can range from hand-operated derricks and winches to overhead cranes and aerial baskets. Hazards associated with **hoisting apparatus** include the following:

- Structural failure or tipping if the apparatus is overloaded or used on windy days
- Material falling on people or property below
- Limited visibility
- Use environment such as power lines.

To **reduce the hazards** associated with hoisting equipment, the following actions should be taken:

- Develop setup standards such as ensuring the setup site is level and away from power lines and that the apparatus is assembled properly.
- Ensure the apparatus is not overloaded. Train operators on both everyday and emergency procedures for using their equipment.

Ropes, chains, and slings are the rigging between the hoist and the load. Hazards associated with this rigging stem from the following:

- Overloading
- Deterioration
- Improper rigging leading to falls

To **reduce these hazards**, the following actions should be taken:

- Store rigging properly, out of sunlight and away from moisture and chemicals that can cause it to deteriorate.
- Inspect rigging regularly to ensure it is not deteriorating or wearing out.
- Follow load capacity charts for rigging to guard against overloading.

Common Hazards and Controls Associated with Crane Operations

Safety Operations Associated with Cranes and Lifting Devices

Construction cranes account for a high percentage of workplace injury and accidents on construction sites. The CHST can implement several strategies to minimize injury from unsafe conditions:

- Before operating, the crane's **hydraulic lift controls** should be inspected and determined to be in proper working order. The **outrigging hydraulic extenders** should be solidly placed and fully extended to guarantee utmost stability.
- The weight of intended loads must be within the crane's **specified load capacity**.
- **Human walking traffic** should be barricaded from the crane's area of operation. At no time should loads be moved over the heads of workers.
- Cranes must be situated and operated at safe distances from **electric power lines** or onsite **distributive power lines** needed for small equipment.
- **Rigging and brakes** must be examined and trial tested before use. Inspect cable, chains, and hook.

Daily Crane Inspection

A daily crane inspection program is an essential part of safe crane operation. The presence of appropriate personal protective equipment (PPE) should be confirmed, and it should be checked that access to the area that the crane will be used is controlled. Check the disconnect switch for correct operation, that the hoist is not loose or broken, that the wire rope is seated properly and not twisted or bent, that the push button controls have no damage, that the controls have the required American Nation Standards Institute (ANSI) tag, that the controls operate as they should (e.g., the crane goes up when that button is triggered), that hooks have no cracks or gouges, that the wire ropes are not frayed, and that the capacity plate is present and visible.

Monthly Crane Inspection

There are additional inspection elements for a monthly crane inspection above the daily requirements. All inspections must be conducted by qualified personnel. Items to be inspected include looking for deformed, cracked, or corroded members; loose bolts and rivets; cracked or worn sheaves and drums; worn bearings, pins, shafts, or gears; and excessive wear on brake system and parts. Check that the load indicators function correctly over the full operating range, and check for excessive wear on chain drive sprockets.

Crane Load Ratings

Manufacturers of cranes design them for a certain load capacity. The load capacity is the maximum that the crane can safely lift; however, it is not good practice to lift approaching the capacity limit unless the crane is designed for that type of use. Cranes are classified as follows:

- Class A: Standby service—crane on standby or used infrequently
- Class B Light service—used at slow speeds and infrequently
- Class C Moderate service—Average 50 percent of rated capacity, five to ten lifts per hour
- Class D Heavy service—Used at 50 percent of rated capacity constantly through the work day
- Class E Severe service—Used approaching capacity 20 times per hour
- Class F Continuous severe service—Used approaching capacity continuously throughout its life

SAFE OPERATION OF MOBILE CRANES

In order to make a safe lift with a **mobile crane**, the operator needs to know the following:

- Whether the vehicle is level.
- Whether the outriggers are extended or retracted.
- If the outriggers are extended, whether they are supported by stable ground.
- Whether the tires are fully inflated.
- The angle of the boom.
- The length of the boom and jibs.
- The positions the boom will be in during the lift.
- The weight of the load.

With this information in hand, the operator can use a load chart to determine if the load is within the structural and stability limits of the crane.

COMMON HAZARDS AND CONTROLS ASSOCIATED WITH MATERIAL HANDLING

Materials handling involves lifting, moving, and placing items, whether manually or with equipment. Equipment frequently used for materials handling includes jacks, hoists, backhoes, escalators, derricks, and cranes. As many as 20-25 percent of disabling work-related injuries involve materials handling. Hazards stemming from materials handling differ depending on the material being handled and on the equipment used. Materials may be heavy, toxic, radioactive, hazardous, or flammable: each involves a different set of hazards. Equipment used may be mobile, such as a forklift, and could hit someone. Other times, items can fall off a hoist or a crane can collapse. Electrically powered equipment poses electricity hazards such as shock. Moving items manually also poses hazards. For example, lifting too heavy a load or using improper lifting techniques can cause sprains or strains.

MATERIALS HANDLING AND STORAGE SAFETY

Storing and moving construction materials requires a great deal of twisting, turning, lifting, pushing, and bending. According to the National Safety Council, twenty percent of worker injuries are related to **back injury**. To minimize such injuries, OSHA recommends the following relatively simple preventatives:

- Workers should seek help when a load is so bulky or heavy that it cannot be safely moved by one person.
- Holders or handles should be attached to loads to make movement steadier.
- Steel-toed shoes and other protective devices should be worn to prevent injuries to feet from dropped materials.
- Gloves are obligatory to prevent cuts and cracks in the skin which invite injury.
- Walkways must be kept free of debris or extraneous stored materials, especially in areas where workers routinely carry and deliver objects needed in construction.
- Materials which are scheduled to be moved or are already moved should be stacked and racked with blocking which will prevent them from collapsing or rolling onto passing employees.

Preventing Accidents During Materials Handling

Developing and enforcing procedures for materials handling can help **reduce accidents**. Procedures should include how to do the following:

- Select the correct equipment for the job.
- Communicate during the materials transfer, whether using hand signals, two-way radios, or other forms of communication.
- Deal with problems that could occur during the materials handling activity.

The procedures should also include step-by-step instructions for actually completing the materials handling activity. Other actions that can help **prevent materials handling accidents** include the following:

- Create a safe materials handling environment with good lighting, wide aisles, and proper ventilation.
- Institute traffic controls to keep lift areas clear of people.
- Regularly maintain, inspect, and repair equipment used for materials handling.
- Train workers to properly use materials handling equipment, hand signals, and rigging.
- Train workers on safe techniques for manually moving materials, such as techniques for lifting heavy loads.
- Eliminate materials handling whenever possible.

OSHA has regulations which apply to construction and materials handling procedures, but employers or contractors set policy. A good management program designed to **limit worker material handling injury** would include:

- Training in the techniques of proper lifting and conditioning (warm-up and stretching) techniques.
- Training in the use of blocking, stacking, and proper storage of materials should be a requirement.
- Training in the observance of weight limits or maximum capacities for motorized or mechanical equipment should be implemented.
- Employers should set policies of minimum physical requirements which must be met by employees who will be occupied in the movement of heavy materials.
- Employees should be trained in and encouraged to use protective clothing and other equipment designed to prevent exposure to harmful substances.

Safe lifting techniques

Training workers on safe lifting techniques is important because improper lifting can cause sprain and strain injuries. As many as 25% of workers' compensation claims are for lower back injuries. To decrease the chances of injury when lifting, you need to do the following:

- Check the weight of the item to be lifted. If it is higher than the RWL (recommended weight limit), take steps to lighten the load, whether by using equipment to pick up the item or by splitting the load into two or more loads.
- Ensure the floor is not slippery.
- Spread your feet for a more stable stance.
- Keep your back straight.
- Hold the load close to your body.
- If an item is on the floor, bend down, grasp it firmly, and then stand up slowly and steadily.

RWL and LI: RWL stands for **recommended weight limit**. The RWL is the weight that healthy workers could lift for up to eight hours without causing musculoskeletal injuries. To calculate the RWL, you need to multiply LC (load constant) x HM (horizontal multiplier) x VM (vertical multiplier) x DM (distance multiplier) x AM (asymmetric multiplier) x FM (frequency multiplier) x CM (coupling multiplier). LI stands for **lifting index**. It measures the physical stress associated with lifting. As the LI goes up, the chance of injury also goes up. To calculate LI, you divide the load weight by the RWL. It is important to know the RWL and the LI because they can help you in the following ways:

- Identify tasks where injuries are more likely to occur.
- Help you develop procedures for safe lifting tasks.
- Identify which lifting tasks need to be redesigned first (generally those with higher LIs).

Ultimately, knowing the RWL and LI can lead to safer lifting procedures and fewer injuries. In order to determine the RWL, you must know the weight of the object. You must also know the following task variables:

- H (**horizontal distance**) – distance of the hands away from the midpoint between the ankles.
- V (**vertical location**) – distance of the hands above the floor.
- D (**vertical travel distance**) – number of inches or centimeters the object is lifted.
- A (**asymmetry angle**) – distance from 0 to 135 degrees that a worker turns during the lift.
- F (**lifting frequency**) – the average number of lifts per minute, measured over a 15-minute time span.
- C (**coupling classification**) – defined as good, fair, or poor depending on the type of grip (such as handles) and the type of container (box or bag; rigid or non-rigid, standard or irregular shape, etc.).

HANDLING OF DRUMS AND CONTAINERS FOR HAZARDOUS MATERIALS

The OSHA regulations in 1910.120 describe the special handling treatment of drums and containers of hazardous substances and contaminated liquids. Prior to the Environmental Protection Act and other legislation designed to protect workers and communities, it was fairly commonplace to store 55-gallon drums by burying. Regulations require that this situation be ameliorated by interventions.

- Whenever possible, drums and containers should be **inspected and tested for contamination** before being moved. If this is impractical, then the containers should be moved to an area which is more hospitable to inspection of contents.
- Currently, there are stringent regulations regarding **labeling and storage** of containers. If an unlabeled or unidentified container is found at a worksite, it should be handled according to existing regulations for the handling of hazardous materials and wastes until the substance is identified.
- Well-planned worksites should be controlled so that **movement of containerized chemicals and substances** is minimal or non-existent.
- Employees occupied in the on-site movement of containers and drums must be **trained** in the potential hazards and informed of the potential for health effects.

Moving and Storing Compressed Gas Cylinders

Compressed gas cylinders are dangerous when dropped, exposed to ignition sources or corroded. Precautions are to be taken when moving and storing cylinders of compressed gas:

- **Valve protection caps** should be placed over all valves. Valve protection caps are not "handles" and should not be used for lifting and moving cylinders.
- Before hoisting cylinders aloft, compressed gas cylinders must be **stabilized** on a cradle, slingboard, or pallet. Magnets or choker slings may not be used for moving gas cylinders.
- When moving cylinders, they should not be **dropped** or permitted to crash into each other violently. The recommended method is to tilt them and roll them on their bottom edges.

Precautions are to be taken when moving cylinders of compressed gas by motor vehicle.

- It is important to recognize that cylinders of **oxygen** must be separated from **gas** cylinders or other combustible materials when moving them by truck.
- If moved and then stored, the oxygen cylinders must be set apart at distance of **20 feet or more**. If this cannot be done, then a **noncombustible barrier** must be erected to a height of at least 5 feet. The barrier must have a fire-resistance rating of at least ½ hour.
- Cylinders moved by motor vehicle must be loaded in a **vertical position**. Unless a special carrier is used, gas regulators must be removed and replaced with valve protection caps before the cylinders are moved.
- Unless transported by a dedicated cylinder truck with racks, a **chain or stabilizing device** should be used to maintain cylinders in an upright position.

Preventive Safety Measures for Workers Using a Conveyor Mechanism to Move Materials

There are specific dangers associated with conveyor mechanisms used to move materials from one area to another:

- Hands, gloves, or clothing may be caught in **rollers or pinch points** as materials are moving. Loose flowing clothing should not be worn when working near conveyors.
- **Emergency buttons or pull cords** set up to halt the conveyor must be conveniently located and accessible to prevent minor accidents from becoming major ones.
- Before a stopped conveyor is restarted, a complete **inspection** of the conveyor should be conducted by safety personnel to determine the origin of the stoppage.
- **Safety nets or barricades** should be constructed to prevent materials from falling from conveyors and striking nearby workers.
- Conveyors are intended to move materials, not **people**. Employees should not sit on or attempt to ride conveyors.

Common Hazards and Controls Associated with Material Storage

Controls for Storing Materials Safely

To store materials safely, the following actions should be taken:

- Use good **housekeeping procedures** so that everything has a place where it belongs.
- Don't **stack** materials too high. Stacks that are too high take little force to be knocked over. In addition, if materials are stacked too high, lower boxes can be crushed and the stack could then tip over.
- Use **crossties** for lumber and bagged materials.

- Use **stepping back procedures** when materials are stacked in several rows.
- Use **retaining walls** to restrain bulk materials.
- Use **racks** to keep items like drums, rolls, and piping from falling.
- Use **protective barriers** to keep industrial trucks from running into racks.
- Design storage areas with **wide enough aisles** for industrial trucks if they will later be used to move the stored materials.
- Use **netting** to catch items that may fall from overhead storage areas.

STORING FLAMMABLE OR COMBUSTIBLE MATERIALS WITHIN A CONSTRUCTION AREA

The use of proper storage containers for flammable or combustible materials cannot be overlooked in the interest of worker safety.

- If combustible or flammable liquids must be stored in an amount more than one gallon, only approved **metal safety cans** may be used.
- Flammable and combustible materials must not be stored on **stairwells, hallways, or any passages** where workers routinely travel.
- If **more than 25 gallons** of flammable or combustible liquids must be stored, the method of containment should be constructed solidly and according to the following specifications:
    - Metal containers are preferable. If wooden cabinets are used, they must be constructed of exterior plywood no less than one inch thick. Joints must be rebated and fastened with wood screws to prevent separation. Hinges must be strong and secure and the cabinets should be clearly labeled as "flammable" or "combustible" and painted with a fire-resistant paint.

STORAGE AND HANDLING OF FLAMMABLE MATERIALS

Flammable liquids must be stored in approved flammable containers; for example, flammable liquids will generally be supplied in glass bottles or metal drums, not in plastic containers. If a secondary container is to be used (e.g., to transport a small quantity of gasoline), the container must be approved to hold flammable liquids. Flammable liquids must be stored in approved flammable liquid storage cabinets or in segregated rooms that have separate ventilation and whose walls meet fire resistance ratings. Aisles must be maintained at 3 feet in width, and egress routes must be kept clear in areas where flammable liquids are stored.

STORING FLAMMABLE AND COMBUSTIBLE MATERIALS INDOORS

When storing flammable or combustible materials indoors, the following actions should be taken: Store only small amounts in occupied areas or buildings; larger amounts need to be stored in separate facilities. Follow **National Fire Protection Association (NFPA) standards** when designing storerooms for flammable and combustible materials, including standards for ventilation, static electricity grounding systems, explosion-proof light switches and fixtures, self-closing doors with raised sills, signage, and floor contours. Follow NFPA standards for storage cabinets designed to hold small quantities of flammable and combustible items. When transferring flammables from one container to another, ensure that the containers are touching each other or are connected to a grounding rod or line. Store flammable liquids in closed containers. When dispensing flammable liquids from drums, use gravity or suction pumps rather than pressurizing the drum. Use safety cans to move flammable liquids from their storage area to their point of use. Use plunger cans if you need to wet cleaning cloths with a flammable liquid. Store cloths that have been contaminated by flammable liquids in a small self-closing container that you empty regularly.

Requirements Applying to Specifications for Concrete Mixing, Storage, and Moving

OSHA regulations regarding concrete storage, mixing, and moving operations are designed to protect workers from injuries stemming from uncontrolled flows of concrete:

- Containers and bulk storage bins for concrete should be equipped with **tapered bottoms** to control the volume being poured.
- Concrete silos used to store bulk concrete must be equipped with **mechanical or pneumatic drives** to start and control the flow of material.
- Concrete silos and bulk storage bins should have **prohibited entry systems** which do not permit employees to access them unless the operating systems have been shut down, tagged, or equipped with "lockouts" which prevent operation.
- Concrete "buckets" must be equipped with mechanically operated valves which have **safety blocks** to prevent accidental dumping of materials.

Storage Stacking Requirements for Masonry Blocks, Bricks, and Concrete

OSHA regulations in **1926.250** offer the provisions for stacking and storing various types of material: bagged materials, bricks, masonry blocks, and both used and new lumber. The provisions of 1926.250 state:

- **Bagged materials** like concrete, lime, fertilizers, and feed should be stored and stacked by stepping back the layers and cross-keying. Cross-keying refers to the practice of changing the direction of placing the bagged materials and interlocking them with layers below.
- **Bricks** must not be stacked more than seven feet in height. For heights of more than 4 feet, the brick should be tapered back and way at a slope of 2 inches per foot of height.
- **Masonry blocks** stacked higher than six feet must be tapered back a length of one-half block for every layer above the six-foot level.

Control and Storage Measures for Chemicals

Chemicals should be stored separately according to key physical and/or chemical properties. They may be stored alphabetically, but only within a designated group. For example, oxidizing agents should NEVER be stored near organic solvents. Additionally, some chemicals are limited in the amounts that can be stored on-site at any one time. Chemicals must be stored in appropriate containers. For example, some organic solvents will dissolve some types of plastic. Alkaline solutions often etch glass containers. The containers must be appropriately sealed to minimize escape of vapors. Such chemicals are often stored in a cabinet that is vented into the exhaust system. Bottles of chemicals must be scrupulously labeled with the name as provided on the SDS sheet, potential hazards, and the contact information of the manufacturer. Caution signs should be posted at entrances to chemical storage rooms that warn of the particular hazard(s) present. Finally, chemicals must be in a secure location at the proper temperature and pressure needed to ensure stability.

Control and Storage Measures for Corrosive Materials

Corrosive materials are those that can dissolve metal and severely damage human tissue upon contact. They are usually of either a very high pH (greater than 12) or low pH (less than 2) but can also be other types of chemicals. Because they can dissolve metal, they should be stored in plastic containers. Corrosive materials must be stored away from flammable materials and incompatible metals. Acids and bases should be stored separately from one another due to the risk of exothermic (heat-producing) reactions if they are accidentally mixed. If possible, they should be stored in an enclosed cabinet or room with ventilation that does not go into the work area.

INCOMPATIBILITY

Incompatibility refers to one or more chemicals that when mixed together, create a hazardous reaction. A hazardous reaction can refer to generation of heat (to the point of starting a fire), creation of a hazardous gas (e.g., hydrogen gas, which is flammable), or creation of a toxic gas. Common incompatibilities are strong acids and strong bases (heat and hydrogen gas), ammonia and bleach (toxic fumes), and strong acids with oxidizers. Incompatible chemicals must be stored in well-ventilated areas well apart from one another to ensure they do not mix accidentally.

HAZARDOUS MATERIALS CONTAINMENT

Hazardous materials containment refers to properly containing hazardous materials in the event of an unexpected spill. Considerations should be made when designing containment as follows:

- Containment capacity: What should the containment capacity be? The design criteria are to contain 110 percent of the largest container or 10 percent of the aggregate volume stored. In addition, if the area is subject to storm water, containment capacity must be provided to contain the storm water also.
- Compatibility: Ensure that the materials of construction of the containment device are compatible with the hazardous materials being contained.
- Inspection frequency: Plans should be made to inspect the containment for spilled material on a regular basis. A corrective action process should be put in place to address any deficiencies discovered.
- Emergency planning: What will be the procedure if material is spilled in the containment? Who will be notified, and how will the material be cleaned up?

If the spill is small and the hazards are known, the spill can generally be cleaned up by personnel using appropriate personal protection gear and methods specific to the chemical involved. Exceptions include mercury and radioactive substances. Spills can typically be handled using specifically designed adsorption spill kits or via neutralization. Powdered neutralizers should be applied to the spill from the outside to the inside of the spill in a circular manner. Depending on the chemical, the neutralized material should be placed in a carefully labeled container for appropriate disposal. All sources of ignition must be controlled if the substance is flammable. At least two people should be present in case the person cleaning the spill needs emergency assistance. If large amounts of chemicals have spilled or there is a possibility the spill can result in an uncontrolled release of a hazardous substance, the designated HAZMAT team must be summoned.

CONTROLS FOR PREVENTING OR REDUCING SEVERITY OF EXPLOSIONS

The exact controls used to prevent an explosion or reduce the severity of an explosion will change depending on the materials and use environment. However, some general guidelines include the following:

- Limit the amount of explosive material stored in any one area.
- If large amounts of explosive materials must be stored, they should be in a remote area.
- Regularly clean areas where explosives are stored so dust will not accumulate.
- Eliminate sources of ignition such as lighters, moving belts, and electrical equipment.
- Store fuels and oxidizers in separate locations.
- Install extinguishing and suppression systems to put out fires before an explosion can occur.
- Use vents in any containers where explosive mixtures could lead to explosions.
- Use distance and barriers to separate explosive materials from each other and from populated areas.
- Train anyone handling, using, and distributing explosives.

COMMON HAZARDS AND CONTROLS ASSOCIATED WITH HOUSEKEEPING

HOUSECLEANING, HOUSEKEEPING, AND SANITATION

Housecleaning refers to the process of cleaning an area: sweeping, wiping surfaces, throwing away trash, etc. **Housekeeping** means putting things away where they belong. Every tool, piece of equipment, and material should have a designated storage area. Hazardous materials should have special storage areas designed specifically for them. **Sanitation** means ensuring the facility is clean and germ-free.

It includes such issues as the following:

- Safe drinking water.
- Clean working toilets.
- Clean areas to prepare and eat food.
- Insect and rodent control.

Housecleaning, housekeeping, and sanitation can all reduce hazards. For example, putting equipment where it belongs means that no one is going to trip over that item and the item cannot fall on anyone. Providing a sanitary environment also reduces disease transmission and can lessen exposure to hazardous substances.

Effective housekeeping is important from both environmental and safety standpoints. Poor housekeeping can increase the volume of waste generated by causing spills and by accumulating out-of-date materials that must be discarded. From a safety standpoint, poor housekeeping can create slip, trip, and fall hazards; can block emergency exits; and can block emergency equipment such as fire extinguishers. Setting **housekeeping policies** and enforcing them, along with employee training and an inspection program, are the key elements of an effective housekeeping program. Effective housekeeping begins with only necessary materials in the work area, with everything having a designated location, and with clear communication of these guidelines (all elements of a 5S program). Employees should be trained to clean up as they go along throughout their shift and should set aside time at the end of shift to clean up completely. All tools should be replaced into storage lockers and containers returned to storage locations. Surfaces should be kept clean. Effective housekeeping promotes a safe workplace and increases job satisfaction because it is more pleasant to work in an orderly environment.

CONTROLLING HAZARDOUS CHEMICALS

Part of housekeeping is removing dust and cleaning up spills. **Hazardous dust** needs to be regularly vacuumed from surfaces so that it will not become airborne. A vacuum that traps the contaminants must be used. Materials can become airborne when they are loaded, unloaded, and transferred to other containers. Transferring within a closed transfer or exhaust system can protect workers from being exposed to airborne dust and vapors. For **liquids**, it is also helpful to use drip pans or containers to collect overfill spills and leaks. Leak detection programs can include both automatic sensors and regular visual inspections of valves and pipes. The sensors can trigger alarms or even shut down a process. Repairing leaks quickly minimizes any potential exposures. Workers and supervisors who use hazardous chemicals need to receive training on what hazards they face and how to protect themselves. This training will help them stay safe and is also required by OSHA standards and by law in some states.

SAFETY HAZARDS ASSOCIATED WITH POOR HOUSEKEEPING

There are many safety hazards associated with poor housekeeping. There is a potential for slips, trips, and falls if there are objects lying on the floor such as boxes, tools, cords, and equipment.

Material spilled on the floor also increases the chance of slipping, whether the material is oil, water, or a solid substance. Poor housekeeping and management of oily rags and other flammable materials is a fire hazard if the materials are not properly stored or disposed of. Materials that are precariously stacked overhead are a hazard to employees, who can be struck by falling objects. Improper storage and handling of hazardous materials can result in exposure to harmful organic vapors. Improperly managed containers also pose a spill hazard. Good housekeeping practices contribute to a safe work environment, are more pleasant to work in, increase employee morale, and contribute to increased productivity because employees are able to find objects they need when they need them.

Importance of housekeeping for a safe workplace

Proper housekeeping and cleanliness are key contributors to a safe workplace. Good housekeeping diminishes slip, trip, and fall hazards by ensuring that tools and work items are put away when not in use. Good housekeeping practices clean up spilled materials promptly, getting them out of the aisles so that employees won't slip on them. Good housekeeping practices help to minimize waste and assist in proper dust control. Proper waste management and storage practices can also minimize fire hazards. Good housekeeping can also help manage tools appropriately and ensure they are not damaged by improper storage practices.

Materials given special consideration regarding fire hazards

Housekeeping is important in controlling fire hazards, especially storage and handling of rags and wipes soaked in flammable materials. If one accumulates rags and wipes soaked with solvents or similar materials, they should be stored in metal canisters rated for flammable rag storage and emptied daily into suitable containers. If rags and wipes soaked with flammable solvents are stored in piles outside of the metal canister, they can build up heat and create conditions favorable to spontaneous combustion. It is also important to store flammable liquids in an orderly manner in cabinets or rooms rated for flammable chemicals.

Contribution of poor housekeeping to fire hazards

Poor housekeeping can increase fire hazards in a number of ways. Poor housekeeping can lead to accumulation of oily or solvent-laden rags and wipes, which are a combustible hazard. Poor housekeeping will also contribute to a higher level of combustible debris lying around, such as boxes, pallets, and packaging material. Poor housekeeping can lead to blocking of paths of egress and exits, which contributes to fire and life safety hazards. In addition, poor housekeeping can lead to accumulation of dust on surfaces and in electrical panels, which can be combustible or cause a short in an electrical panel that leads to a fire.

REQUIREMENTS FOR MAINTAINING CLEARANCE AROUND ELECTRICAL PANELS AND FIRE EXTINGUISHERS

Occupational Safety and Health Administration (OSHA) regulations do require that electrical panels have a minimum 30 inches of clearance around them for voltages between 120 to 250 volts. Higher voltage panels require greater clearances (29 CFR 1910.403). In most work areas, lines are painted on the floor to indicate the clearance required as a means of reminding employees not to store materials in these areas. In contrast, there is no specific requirement for a clearance around fire extinguishers; however, OSHA regulations do state that fire extinguishers must remain accessible at all times and that employees must not have to travel more than 75 feet to reach a fire extinguisher if they are working in an area with potentially combustible materials (29 CFR 1910.157[c][1]).

Common Hazards and Controls Associated with Powder Actuated Tools

A powder-actuated tool is a forceful **nail gun** that uses a small amount of a **chemical propellant** to force the nail or other fastener into the substrate. The mechanism is somewhat similar to the operation of a firearm. Powder-actuated tools use a cartridge with the fastener loaded in. They can be of either high or low velocity; however, the difference in velocity does not imply a difference in the hazard posed by this equipment. Various types of fasteners are available for use with these tools (such as nails or threaded studs), and these fasteners can be driven into steel or concrete.

Powder-actuated tools deliver sharp fasteners such as nails at high velocity and force. Therefore, they pose an extreme **puncture hazard**. The force and mode of operation also pose a **noise exposure hazard**. There is also a risk of **flying objects** if the fastener is not properly delivered into the substrate. They pose a **vibrational hazard** from the vibration of the tool. For these reasons, eye and hearing protection must be worn when using these tools, along with foot protection. Specific training in the use of these tools must be completed and documented.

Guards and Controls for Powder-Actuated Tools

Powder-actuated tools should be **inspected** prior to each use. The manufacturer's literature should be reviewed to determine the engineered safety features of the tool. Ensure that the tool is equipped with an **ignition safety feature** so that it doesn't fire when pressed against a surface. Ensure that it is equipped with a **drop safety feature** so that it will not fire if dropped. Inspect the fastener cartridges before use to ensure they are intact and not compromised. There are no particular guards required, but warning signs must be on the tool case, on the tool itself, and posted in the work area stating, "Powder-Actuated Tool in Use."

Materials to Avoid When Using Powder-Actuated Tools

Powder-actuated tools drive sharp objects at high force to fasten two pieces together. As such, any material that is **brittle** or will **shatter** should not be fastened with a powder-actuated tool. Examples of these types of material are cast iron, marble, glazed ceramic tiles, glass blocks, slate and other types of natural stone, and clay bricks. If powder-actuated tools will be used to fasten into **concrete**, the concrete must have appropriate **compressive strength** to ensure proper fastening. Concrete of compressive strength greater than 8,500 psi is not suitable for powder-actuated fasteners.

Common Hazards and Controls Associated with Hand and Power Tools

Hazards and Injuries Associated with Tools and Machines

Hazards associated with tools and machines include the following:

- Being struck by a tool, machine, or machine part.
- Being struck by flying debris from materials the tool or machine is acting on, such as concrete chips.
- Getting caught in a machine or tool.

Less direct hazards also exist. For example, if a machine requires repeated motion, cumulative stress disorders can result. For powered tools, electrical hazards are also present. Tools and machines are a major source of injury, including the following:

- Cuts
- Abrasions
- Puncture wounds

- Tissue tears
- Crushing injuries
- Fractures
- Carpal tunnel syndrome
- Bursitis
- Tendonitis

As many as 8% of injuries involving lost time are related to **hand tools**.

Hand Tool Inspections

The term "hand tools" refers to handheld tools that are not powered. This includes screwdrivers, hammers, axes, scissors, box cutters, and pliers. All tools should be inspected before every use to ensure they are in good operational order. One should examine hand tools to ensure the metal is not cracked or thinning to the point that it may break during use. One should also check the handles to ensure they are firmly seated and that any grip coverings are in place and not loose. Tools that are comprised of several parts should be checked to make sure that the parts are securely attached.

Power Tool Inspections

Power tools refer to handheld tools powered by electricity, batteries, compressed air, or internal combustion engines. These tools include power saws, drills, nail guns, grinders, cutting tools, sanders, and riveters. Some inspection elements are common to other handheld tools, such as checking that the metal parts are not cracked or worn and that all parts are firmly seated and not loose. The electrical cord should be inspected to make sure it is not worn or the insulation frayed. The electrical prongs should be inspected to make sure they are not bent or missing (especially the round grounding prong). If there are cutting blades in the tool, they should be inspected to make sure they are not too worn or needing sharpening; dull blades are unsafe due to the extra force required to use them that may cause one's grip to slip.

Pneumatic Power Tools

The development of pneumatic hoses to power construction tools is a relatively recent innovation in the construction industry and has presented a new potential for safety problems. OSHA guidelines in 1926.302 recommend:

- Pneumatic power tools should be locked and secured in place to prevent the tool from being forcefully ejected from the power source. Safety clips should be used for this purpose.
- Pneumatic power nailers and fasteners operating at more than 100 psi should have built-in safety mechanisms which prevent triggering unless the muzzle is pressed against the desired surface.
- Compressed air at more than 30 psi. cannot be used for cleaning purposes.
- Pneumatic hoses more than ½ inch inside diameter must be equipped with safety devices which reduce pressure if hose lines are ruptured.

Management System to Ensure Tools Are Inspected Regularly

An effective management system to ensure tools are inspected regularly must first identify what types of tools are on hand and in which locations and departments. A schedule of inspection and a responsibility matrix (by job title) must be developed. Finally, the system must include inspection checklists to document findings and follow up corrective actions. Calendar systems can be used to schedule inspections and ensure deficiencies are corrected in a timely fashion. Coupled with routine inspections, random inspections of tools in use on the floor are important as cross-checks to ensure the inspection program is operating properly.

COMMON HAZARDS AND CONTROLS ASSOCIATED WITH ASBESTOS EXPOSURE

ASBESTOS IN THE WORKPLACE

Asbestos is the common name for a group of minerals like chrysotile, amosite, crocidolite, tremolite, and other mineral compounds used and found in the construction workplace:

- Surfacing materials contain asbestos substances. These may be sprayed, brushed, or troweled onto surfaces. Wallboard, floor tiles, and mastics are all commonly found to contain asbestos materials.
- Asbestos products are found in thermal insulation piping, tanks, ducts, boilers, and other heating and cooling equipment structures.
- There are four classes of asbestos-related construction activity listed according to the degree of hazard and contact. These categories range from Class I to Class IV.
- Each exposure class has a specific individual protocol for handling ACM (asbestos containing material). OSHA requires that the process of handling ACM be managed by "competent persons". Competent persons are those who have been certified in asbestos handling procedures.

Employers must control worker exposure to asbestos to the extent possible, monitoring air quality, providing protective equipment, and medical testing for workers who are subject to Class I, II, or III work activities for 30 days or more.

- Employers must retain training records up to one year extending from the last day of a worker's employment.
- Employers must retain medical records during the period the worker is employed and for 30 years afterward.
- Employers must establish hygienic decontamination areas for workers.
- Employers must establish limiting control zones which permit access only to authorized persons. Eating, drinking, smoking, applying cosmetics shall not be permitted in restricted areas to prevent ingestion or other contamination.

APPROACHES FOR DEALING WITH ASBESTOS

The first decision to make regarding **asbestos in the workplace** is whether it should be left alone or abated. If the asbestos is not friable and is not in an area where it will be disturbed, it may be safer to leave it alone. However, if the asbestos is friable or is, for example, in an area scheduled for renovation, then the **asbestos-containing material (ACM)** must be removed, enclosed, or encapsulated. If the decision is to leave the asbestos alone, an asbestos-monitoring program must be put in place to track whether the condition of the asbestos changes. Any work done with asbestos needs to be performed by trained and licensed personnel wearing high efficiency respirators and disposable personal protective equipment that covers the entire body. *Removing* ACM is the most expensive choice, but once the ACM is removed, no further work or monitoring is needed. When ACM is removed, the area where it is located must first be enclosed in tough plastic walls. That area then must be ventilated by HEPA filtered negative air machines. The ACM is covered by a liquid solution that keeps asbestos fibers in place and is then put in leak proof containers for disposal. If ACM can be encapsulated, it is sprayed with a sealant that binds the asbestos fibers together. The sealant hardens into a tough skin. ACM can also be *enclosed*. This involves placing airtight walls around the ACM. Warning signs need to be placed on the walls and their location needs to be marked on the building plans. If ACM is enclosed or *encapsulated*, an asbestos-monitoring program needs to be set up to track whether the condition of the asbestos changes.

Common Hazards and Controls Associated with Lead Exposure

Lead Contamination in Construction Sites

Though the use of lead-based paint in housing construction was banned in the 1970s, there are still a surprising number of construction processes through which a worker may be exposed. Workers at the highest risk of lead exposure are those involved in:

- Abrasive blasting has the capacity to put lead in the air.
- Welding, cutting, and burning on steel structures creates lead particles.
- Careless or accidental burning of materials containing lead.
- Using lead-containing mortar.
- Power tool cleaning without dust collection systems.
- Rivet busting.
- Cleanup activities where dry expendable abrasives are used.
- Movement and removal of abrasive blasting enclosures.
- Manual dry scraping and sanding.
- Manual demolition of structures.
- Heat-gun applications.
- Power tool cleaning with dust collection systems.

OSHA PEL Standards Which Apply to Worker Exposure to Lead in the Atmosphere

The PEL or Permissible Exposure Limit for airborne lead particles is 50 micrograms of lead per cubic meter of air expressed a 50 µg/m³ per eight-hour period.

- The standard PEL for lead is based on a TWA or time-weighted average.
- A daily exposure may exceed 50 µg/m³ if the worker hours are averaged so as not to exceed the PELs for an eight-hour period.
- If worker exposure to lead will be maintained for a ten-hour workday, the PELs are reduced to 40 µg/m³ according to a formula.

The purpose of the revised formulas for workdays exceeding eight hours is to recognize true workplace demands.

Increasing Protections Against Lead in the Workplace

Hygiene and access to washing facilities are an important part of an employer's lead monitoring and control program. The objective of any program is to minimize the amount of lead uptake through ingestion, inhalation, or through skin contact.

- The employer must maintain separate areas for the consumption of food or beverages. The same applies for areas where tobacco products are used. Cosmetics should not be applied in an area where lead is present since these absorb particles of lead which then come in contact with skin.
- When airborne lead exposure exceeds the PEL, the employer must provide clothing changing areas and showers. Disposable protective clothing must be isolated and disposed of appropriately. It is important to prevent cross-contamination between PPE and clothing.

It is both vital and mandatory to monitor, train, and educate workers as to the dangers of lead exposure in the workplace. The following measures should be taken to protect the worker and his family:

- Remove protective and disposable work clothing at the end of the work shift. Avoid cross-contamination with street clothing.
- Wash hands, face, and any other exposed body parts before changing into street clothing. A shower is the best means of washing lead particles from the body if such facilities are available.
- Wash hands, face, arms and other exposed body parts prior to eating, drinking, smoking or applying cosmetics.
- Store street clothes in an area which is safe from contamination from lead dust.

COMMON HAZARDS AND CONTROLS ASSOCIATED WITH NOISE EXPOSURE

The primary hazard of noise is **hearing loss**. Noise-induced hearing loss is related to the amount of time a person is exposed to the noise, the decibel level, the frequency, and whether the noise is continuous or intermittent. Types of noise-induced hearing loss include the following:

- **Temporary threshold shift**, which is caused by a short exposure to loud noise.
- **Permanent threshold shift**, which is caused by continuous exposure to noise.
- **Temporary or permanent acoustic trauma** caused by a loud noise, as from an explosion.

In addition to hearing loss, noise can also interfere with **communication**. Noise can make it difficult to hear warnings and sirens and even to communicate normally. It also interferes with learning, causes a startle response and other physiological problems such as high blood pressure and ulcers, and makes people irritable and frustrated.

A **baseline audiogram** is a valid audiogram done after a quiet period and used as a comparison for future audiograms to see if hearing thresholds have changed. **Decibel** (dB) is a unit that defines the intensity of sound. **Hazardous noise** is any sound that can cause permanent hearing loss in a specified population. OSHA has established allowable daily amounts of noise that workers can safely be exposed to. The **noise dose** is the percentage of this daily exposure that a particular sound meets. A **noise-induced hearing loss** is any sensorineural hearing loss that can be linked to noise and for which no other cause can be identified. **Threshold of hearing** is one dBA. This is the weakest sound that a healthy human can hear in a quiet setting. **Threshold of pain** is 140 dBA. This is the maximum level of sound that a human can hear without pain.

> **Review Video: Noise Hazards in Occupational Health**
> Visit mometrix.com/academy and enter code: 279189

CONTINUOUS AND IMPACT NOISE

Since decibel levels of noise and duration of exposure are primary factors which indicate the type and degree of noise level protection required for construction site workers, it is important to know the difference between "impact" and "continuous" noises as OSHA regulations describe them. When sound levels are measured at a construction site, permissible worker exposures are clearly described in OSHA tables.

- For a full 8-hour work-day, a worker must not be exposed to noise greater than a 85-decibel level. As the level of sound begins to exceed that level, so does the level of permitted exposure begin to decline. If continuous sound levels rise to 115 decibels, a worker is permitted no more than 15 minutes of exposure, according to the OSHA table.
- Impact noise levels are short duration sounds emitting from such equipment as nail guns, air hammers, or punch press equipment. A worker must not be exposed to decibel levels which exceed 140 dB of impact noise.

MEASURING NOISE LEVELS IN A WORK AREA

There are two basic types of noise metering instruments available for onsite noise measurement to ensure compliance with OSHA noise regulations:

- General area noise meters are designed to measure the noise level in the areas where they have been placed. These instruments may be moved to other areas but have the disadvantage of providing only generalized data. Workers are often moving through a construction site where the noise levels vary according to the types of activity and machinery being used. A generalized metering scheme provides little data about the individual exposure record.
- Personal Noise Meter devices are designed to be worn by working personnel. Personal noise meters obtain individual readings. This type of noise metering has a practical advantage since it is common for workers to move through various work areas with different levels of noise production.

Whatever type of device is used, it is important to follow proper handling procedures:

- Check batteries prior to use. Batteries should be removed from any sound meter which will be stored for more than 5 days.
- Use a windscreen to protect the microphone in areas where dust and airborne debris may be lifted into the air by excavating machinery or equipment.
- Never cover the microphone pickup with plastic or other material which will distort reception and invalidate the readings.
- Calibrate the meter periodically using the manufacturer recommended calibration device.

DETERMINING ADEQUATE LEVEL OF HEARING PROTECTION

The OSHA hearing conservation standard establishes a permissible exposure limit to noise at 85 decibels (dBA). Exposures above 85 dBA require employees to wear hearing protection in the form of muffs or earplugs. Hearing protection devices have a Noise Reduction Rating provided by the manufacturer. Theoretically, wearing the device reduces the noise exposure by that number of decibels. In practice, one applies a safety factor. This calculation requires the "A-weighted" sound-level readings from a noise meter (the A-level readings are expressed as "dBA" and represent the exposure as heard by the human ear, with very high and low frequencies screened out). The calculated employee exposure wearing the hearing protection is as follows:

Estimated Exposure (dBA) = TWA (dBA) - (NRR - 7)

If C-level noise readings are available, the Noise Reduction Rating can be subtracted directly from the dBC reading to determine estimated employee exposure.

HEARING PROTECTION FOR WORKERS EXPOSED TO NOISE LEVELS EXCEEDING 85 DB

All employees exposed to an 8-hr. time-weighted average of 85 dB noise or greater should be provided with appropriate hearing protection at no cost to them.

- Employers must ensure that hearing protection is being worn by workers subject to high decibel sound levels (85 dB) during an 8-hour time-weighted period.
- Hearing protection is required of any employee or worker who has experienced a threshold shift in hearing level.
- Employees must be allowed to select the type of hearing protectors which they need from a variety of approved protectors.
- It is the employer's responsibility to reevaluate and inspect hearing protection effectiveness when changes in work activity expose the worker to greater levels of sound.

CONTROL METHODS FOR EXPOSURE TO EXCESSIVE LEVELS OF NOISE

Engineering controls are the first consideration for control of excessive exposure to noise. Sound-dampening foam products can be used to line enclosures and dampen noise from machinery. Where noise can't be avoided, its effects can be reduced by grouping and enclosing noisy processes in a soundproof area so that people working in other areas are not bothered. Design features that can help reduce noise include the following:

- Controlling the direction of the source.
- Reducing flow rates.
- Reducing driving forces.
- Controlling vibrating surfaces.
- Using barriers and shields.
- Building with sound-absorbing materials.

In addition to controlling the noise source itself, you can also protect the workers by requiring **protection** such as earplugs or muffs.

Earplugs are soft foam plugs that expand to block noise from entering the ear canal. Earmuffs surround the entire ear and prevent noise from entering. Either may be used to achieve the proper level of noise reduction. Devices are assigned a Noise Reduction Rating (NRR) by the manufacturer after testing to determine how effectively they block sound.

COMMON HAZARDS AND CONTROLS ASSOCIATED WITH RADIATION EXPOSURE

There are three types of radiation: alpha, beta and gamma. Proper shielding is the best protection from the harmful effect of radiation.

Alpha – Alpha radiation is made up of small, positively charged particles. Due to an alpha particle's size and characteristics, it cannot travel great distances and is easily stopped by clothing, gloves, or a piece of paper. Care must be taken not to breathe alpha particles into the lungs, as they will damage the lungs and potentially cause cancer.

Beta – Beta radiation has a higher energy level than Alpha radiation and therefore a greater ability to penetrate surfaces. However, it can be easily blocked by a layer of aluminum foil or similar material.

Gamma – Gamma radiation has the highest energy level and has the most ability to penetrate the human body. Lead is the most common shielding used for gamma radiation; several centimeters is usually sufficient to shield from gamma radiation. Water can also be used for shielding; nuclear reactors and power plants use water shielding. Several feet of water are needed to shield from gamma radiation.

NON-IONIZING RADIATION

OSHA defines "non-ionizing radiation" as a series of energy waves from electric and magnetic fields. As with other radiation sources, these types of radiation travel at the speed of light. There are several sources of non-ionizing radiation:

- Power lines, electrical equipment, and household electric wiring produce extremely low frequency (ELF) radiation.
- Radio frequency and microwave radiation may be absorbed through the skin and body. At high levels, this type of radiation can cause tissue damage due to the heating of the body. Radio transmitters and cell phones are examples.
- Infrared Radiation is emitted from such sources as heat lamps and lasers that operate from this wavelength.
- Visible Light Radiation can damage eyes and skin if it is too intense. The visible light spectrum is what we normally see with our eyes.
- UV or ultraviolet radiation sources come from welding arcs and UV lasers.

IONIZING RADIATION

Ionizing radiation is radiation that can produce ions when it interacts with atoms and molecules. Types of ionizing radiation include x-rays, alpha particles, beta particles, gamma radiation, and neutrons. Ionizing radiation can come from natural sources, such as cosmic radiation and radioactive soils, and from artificial sources, such as television sets, diagnostic x-rays, and nuclear fuels. Exposure to ionizing radiation damages human cells, especially rapidly developing cells. It is especially dangerous for infants and children who have the most rapidly developing cells. Exposure to high doses of ionizing radiation causes radiation sickness characterized by weakness, sleepiness, stupor, tremors, convulsions, and, eventually, death. Low doses may cause more delayed effects, such as genetic effects, cancers, cataracts, and shortened life span.

PREVENTING DAMAGE CAUSED BY RADIATION

The damage caused by ionizing radiation depends on the type and dose of the radiation, the tissue and organs exposed, and the age of person being exposed. The best way to control potential damage is to limit the amount of radiation people are exposed to by limiting the amount of source material. It is also important to limit the amount of time people are exposed to radiation. Other ways to reduce exposure to ionizing radiation include the following:

- Increasing the distance between people and sources of ionizing radiation.
- Using shielding such as air, hydrogen, and water to protect people from sources of radiation. The material used as a shield depends on the type of radiation.
- Using barriers such as walls and fences to keep people away from sources of radiation.
- Use liners and protective materials to keep contaminated waste from leaching into groundwater.

A key concept in radiation protection is to keep the radiation dose "As Low as Reasonably Achievable (ALARA)". Regardless of whether the radiation is alpha, beta, or gamma particles, both time and distance can be used to reduce the amount of exposure. Work schedules should be arranged so that exposure time is limited. Reducing the time exposed to radiation reduces the absorbed dose in a directly proportional manner. Distance away from the radiation source can also be used as a means of reducing absorbed radiation. As one moves away from the radioactive source, the exposure decreases according to the inverse square of the distance as illustrated by the following formula: intensity is proportional to $1/x^2$ where "x" is the distance.

Safety Controls for Radiation

Warnings need to mark any areas where ionizing radiation is located as well as equipment that uses ionizing radiation. In addition to signs on these areas, flashing lights and audio signals can serve as additional warnings. **Evacuation** is a tool used to remove people from an area where a significant amount of ionizing radiation has been released. **Security procedures** need to be in place to keep sources of ionizing radiation from getting into the wrong hands. Procedures can include physical monitoring, controlled entry and exit, and manifest systems. **Dosimetry** measures people's exposure to ionizing radiation. It is necessary because we cannot see or feel this radiation. People who work with or near ionizing radiation need training about its hazards and how to protect themselves and others. **System design and analysis** can help prevent dangerous exposure to ionizing radiation by anticipating and preventing possible sources of failure.

Radiation Counters

There are several different methods and devices for detecting and determining exposure levels of ionizing radiation. Common devices used to determine exposure to individuals are dosimeters or film badges which are worn by the person. When exposed to ionizing radiation elements within the film react and when developed result in a permanent radiation exposure record. Dosimeters similarly react to the ionizing radiation and register a measurement of the exposure and in the case of a pocket dosimeter, which uses an ionization chamber, can be read immediately. Ionization chambers measure radiation exposure by the conductivity created by interaction of radiation with the enclosed gas. These chambers are used by themselves, in dosimeters, and are modified for use in Geiger-Mueller Counters. Scintillation instruments, which use phosphors or crystals to produce light in reaction to radiation, have been found to be extremely sensitive and useful when detecting low levels of exposure.

Common Hazards and Controls Associated with Silica Exposure

Jobs or Tasks That May Be Associated with Silica Exposure

Silica (chemical formula SiO_2) is a naturally-occurring mineral that is the main ingredient in **glass**. Silica can exist in both crystalline and amorphous forms. **Crystalline silica** is a health hazard because the dust produced from the material forms sharp shards that can be inhaled. They can lodge deep in the lungs and cause **silicosis**, which is a debilitating and incurable disease. In the construction industry, crystalline silica may be encountered in cutting and fabricating granite and other naturally-occurring rocks, in concrete, drywall, bricks, grout, sand, and tile. Any activity using these materials that creates dust creates a potential exposure to crystalline silica that needs to be evaluated.

Elements of a Written Silica Exposure Control Plan

The purpose of an exposure control plan is to identify potential materials and tasks that could expose workers to **crystalline silica dust** and present the methods that will be used to prevent and control that exposure. The written plan must formally assign responsibility to a qualified, competent person to carry out the elements of the exposure control plan. The plan must specify procedures that will be used to control exposures, such as the types of saws that will be used in cutting silica-containing materials to minimize dust. The plan must detail the housekeeping procedures that will be used to clean up dust that potentially contains silica while preventing re-entrainment into the atmosphere (for example, wet sweeping or HEPA filter vacuums). The plan must also account for the medical monitoring program offered to employees to provide chest x-rays and spirometry. The plan must also outline the employee training with regard to silica exposure and controls, and must outline the records that will be maintained in support of implementing the control plan.

ENGINEERING CONTROLS FOR SILICA

Engineering controls for any safety hazard are always preferable to control with personal protective equipment alone. The **OSHA silica exposure rule** mandates engineering controls when using saws to cut material that potentially produces an exposure to silica. The saw must be fitted with an **integrated water delivery system** that wets the material during the cutting process to reduce the quantity of dust produced. The use of engineering controls must be augmented by use of **respiratory protection** if the sawing task will be done for more than four hours per day.

HOUSEKEEPING CONTROLS FOR SILICA

Housekeeping is important to minimize exposure to silica. Best practices for housekeeping are designed to prevent build up, disturbance, and re-entrainment of any dust that may contain silica. Dust should be cleaned up regularly using wet sweeping or vacuuming with a HEPA-filter vacuum. **Wet sweeping** involves misting the area to be cleaned with a light mist of water (for example, from a garden sprayer) and then sweeping up the material. The swept-up material should be stored in a covered container. Vacuums with **HEPA filters** are necessary to ensure that the dust isn't emitted into the atmosphere in the exhaust of the vacuum. Using compressed air to blow dust from surfaces should be expressly forbidden.

RECORDKEEPING REQUIREMENTS OF THE OSHA SILICA RULE

The OSHA silica rule requires employers to keep records to demonstrate compliance. **Exposure assessments** for silica exposure must be kept, showing that the exposure was assessed to determine if it exceeded the action level of 25 µg/m³ (averaged over an eight-hour day). Records of medical monitoring assessments of workers must also be maintained. In the context of the silica exposure rule, **medical monitoring** must be offered, including chest X-rays and lung function tests (spirometry). **Spirometry records** are also required if employees will be required to wear respirators, and records of respirator fit tests and respirator training must also be maintained in the event employees are required to wear respiratory protection.

COMMON HAZARDS AND CONTROLS ASSOCIATED WITH CHEMICAL EXPOSURE

CHST RESPONSIBILITIES IN THE AREA OF HAZARDOUS CHEMICALS

The CHST has responsibility for worker protection from chemical hazards in the workplace. Accident preventive strategies pertaining to hazardous chemicals may include:

- A spill prevention plan. Such a plan should be formulated for each type of chemical containing product used in the workplace. The SDS for that particular product should be reviewed and considered in devising the specific plan for that product. A spill-kit should be placed close to where the chemical will be used in advance.
- Workers in the vicinity of hazardous chemicals and materials should be informed of proper handling techniques and protective measures. This information can be provided through oral instruction and reinforced by appropriate signage. Language and instructions should be clear and understandable.
- In addition to verbal directives, workers should be encouraged to read the SDS themselves and to take the recommended personal protection measures. The CHST should emphasize the importance of following SDS requirements for safe storage and disposal of hazardous materials.

Permissible Exposure Limits (PEL)

The major distinction between Permissible Exposure Limits and Threshold Limit Values is TLV's are guidelines set by an organization, while PEL's are regulations set by OSHA. They are both valuable in ensuring a safe workplace, but PEL's have the power of legal enforcement behind them. The PEL is measured and reported as a time weighted average (TWA). The OHS ACT (General Duty Clause) was enacted in 1970 and requires employers to provide a safe working environment, which includes ensuring toxins are maintained below the PEL (or other guideline if OSHA has not determined a PEL for the substance) determined for the chemical.

Threshold Limit Values (TLV)

The American Conference of Governmental Industrial Hygienists, Inc (ACGIH) has developed guidelines, called Threshold Limit Values, some of which have been adopted by OSHA, that estimate the limit that a worker may be exposed to a substance during a standard, 40 hour, workweek without experiencing undesirable effects. The guidelines are based on inhalation data obtained from scientific journal articles. The two most common units are mg/m³ and parts per million (ppm). The formula for converting between the two units is: $ppm = \frac{\frac{mg}{m^3} \times 24.45}{MW}$ where MW references the molecular weight of the substance. FORMULA: MW = Σ(atomic mass)(subscript). The units of MW are gram/mole. Care needs to be taken in interpreting reported TLV values. Some values are averaged over typical 40-hour work week (TWA), some are taken as 15 min (STEL) snapshots, four times a day. The "ceiling" TLV is the maximum exposure at any time for any duration.

Important Terms Related to Hazard Exposure

- **Dose threshold** is the minimum dose of a substance needed to produce a measurable effect. Thresholds are determined, in part, by observing changes in body tissues, growth rates, food intake, and organ weight.
- **Lethal dose** is the amount of a substance that is likely to cause death. The lethal dose of a substance is determined by testing on animals.
- **Lethal concentration** is the amount of an inhaled substance that is likely to cause death.
- **Latency period** is the amount of time between exposure to a chemical and observable effects from that chemical. The effects may be immediate, as with a chemical burn, or delayed, as with cancer.
- **Acute exposure** refers to disease or effects that occur after only one exposure to a chemical while **chronic exposure** refers to disease or effects that occur only after repeated exposure to a chemical.
- **Local effects** are effects from a chemical that injure the skin, eyes, or respiratory system.
- **Systemic effects** are effects from a chemical that damage organs or biological functioning.
- **Carcinogens** are substances that produce cancer in animals or humans.
- **Mutagens** are substances that change the genetic structure of an animal or human, affecting the health of future generations.
- **Teratogens** are substances that cause a fetus to be malformed.
- **Irritants**, including **nuisance dusts**, are substances that irritate the skin, eyes, and the inner linings of the nose, throat, mouth, and upper respiratory tract. These substances do not do any irreversible damage.
- **Asphyxiants** are materials that displace oxygen, interfering with breathing and oxygen transport in the blood.

- **Narcotics** and **anesthetics** keep the central nervous system from operating correctly. Used at the right dose, these substances cause no serious or irreversible effects. However, if too much of the substance is used, unconsciousness or death can result.

ENGINEERING CONTROLS FOR PROTECTING PEOPLE FROM EXPOSURE TO HAZARDOUS CHEMICALS

Engineering controls include substitution, isolation, and ventilation. **Substitution** means substituting a non-hazardous or less hazardous material for a hazardous material. For the substitution to be effective, the new material needs to work as effectively as the old material. **Isolation** means creating a barrier between workers and the source of contamination. This barrier could be a glove box that encloses the hazardous material or a separate enclosure that workers can only access remotely. Isolation can also mean separating hazardous and non-hazardous processes. **Ventilation** protects workers from airborne contaminants. With general ventilation, fresh air replaces or dilutes contaminated air. Local exhaust ventilation captures contaminants before they reach people and moves the contaminants to a safe area for treatment.

CHEMICAL SUBSTITUTION

Substitution of safer chemicals for more toxic or dangerous chemicals should always be explored as part of harm reduction. A safer chemical is one that is less toxic to workers or poses fewer physical hazards such as flammability. A safer chemical can also result in a product that is safer for use by the end user or public. Using safer chemicals can save a company money on personal protective equipment (not required if employees are not exposed). It can also greatly decrease associated hazardous waste disposal costs if toxic waste does not have to be disposed of.

VENTILATION

Ventilation refers to the movement of air and exchange of fresh air for trapped air. Ventilation is important in all indoor environments but is especially important if toxic or harmful vapors are in the work area. For example, in a laboratory setting, ventilation hoods are installed, and workers are instructed to use all volatile solvents inside the ventilation hood. The ventilation hood is equipped with a fan to pull air out of the hood and exhaust it outside. This keeps the vapors out of the work area and away from the breathing zone of the worker. In industrial environments, ventilation hoods operate similarly and evacuate vapors to the exterior of the building, either with capture technology or without depending on the contaminant. The American Conference of Governmental Industrial Hygienists (ACGIH) publishes "Industrial Ventilation: A Manual of Recommended Practice for Design, which outlines ventilation recommendations for types of chemical atmospheres and work environments and is the standard reference text.

ADMINISTRATIVE CONTROLS FOR PROTECTING PEOPLE FROM EXPOSURE TO HAZARDOUS CHEMICALS

One work modification that can help protect workers from **chemical hazards** involves reducing an individual worker's *exposure time* to hazardous chemicals during any one work shift. For example, workers can share an activity or task so that each worker remains below the exposure limit. When it comes to chemical hazards, one aspect of *personal hygiene* refers to cleansing any contaminated skin whether the skin was contaminated during regular work tasks or because of an accident or spill. Procedures should be in place to specify which soaps and cleaners to use for which chemicals. There may also be specific washing stations and showers for different purposes. Hygiene may also include washing eyes that become contaminated as well as safe areas for changing clothes and for eating and drinking. *Personal protective equipment* provides a final level of protection against chemical hazards. Such equipment may include protective clothing, eyewear, creams and lotions, and respiratory equipment.

CHEMICAL PROCESS SAFETY MANAGEMENT

Chemical process safety management is an OSHA regulation designed to provide safe operation of highly hazardous chemical processes, such as in refineries and large chemical manufacturing plants. The regulation may be found in 29 CFR 1910.119. The overall required elements must document the process, the chemicals used, and the reactor vessel construction and must conduct a hazard analysis that takes into account potential failure paths and how catastrophic releases and explosions can be prevented. In the case of pressure relief systems, consideration must be made to install and document the type of pressure relief system, and it must be ensured that the valve materials are compatible with the chemicals in the process. Pressure relief systems may be spring-loaded, which provide simple overpressure protection for a tank or piping system. A second type of pressure relief system is a balanced bellows or balanced piston system, which is appropriate to use when superimposed backpressure is variable. Chemical compatibility of tanks, piping, valves, and tools with the reactant and product chemicals must be investigated and documented.

In chemical process safety, detailed process flow diagrams of the entire process must be produced. The process chemistry must be thoroughly documented, and maximum inventories of chemical reactants, intermediates, and products must be accounted for. Safe upper and lower limits of temperature, pressure, and flows must be calculated and documented. Once this is completed, a robust management of change to the process must be implemented that includes a full assessment of potential impacts of process changes. For each change, a documented technical basis for the change must be completed along with an analysis of its impact on employee safety, a list of operational procedures that will need to be altered due to the change, a timeline for the change, and a consideration of authorization needed to implement the change. The hazard assessment portion of chemical process safety can be any accepted method that considers potential failures of the process and how to prevent a resulting catastrophe. Examples of acceptable hazard assessment methods include failure mode analysis, fault tree analysis, and a hazard and operability study.

COMMON HAZARDS AND CONTROLS ASSOCIATED WITH WORKING IN EXTREME TEMPERATURES

WET BULB GLOBE THERMOMETER INDEX

WBGT or Wet Bulb Globe Thermometer index readings are taken with thermometers which measure wet-bulb temperature, dry-bulb temperature, and globe thermometer temperature. The readings obtained are then fed into a formula or index. The formula or index stated below is an index of outdoor WBGT readings:

$WBGT = 0.7\ NWB + 0.2\ GT + 0.1\ DB$

- WBGT refers to the Wet Bulb Globe Temperature Index.
- NWB refers to Natural wet-bulb temperature.
- DB refers to dry-bulb temperature ("regular" air temperature)
- GT refers to globe thermometer temperature.

Thermometers used for these measurements should be placed upon a stand which permits free air-flow. Placement should be representative of real worksite conditions.

Heat is measured in calorie units but the impact of heat upon the human body can depend on a variety of factors like humidity, direct sunlight, and even the metabolic heat caused by work activity. The WBGT measures:

- Globe temperature: This refers to the temperature inside a blackened, hollow globe. It measures simple radiant heat.
- Natural wet bulb (NWB) temperature: This measures the effects of evaporation and convection. In other words, the amount of air movement or convection over a human body contributes to its stress protections. NWB measurements employ a "wet sensor," essentially a wet cotton wick.
- Dry bulb (DB) temperature is measured by a thermal sensor, such as an ordinary mercury-in-glass thermometer, that is shielded from direct radiant energy sources.

HEALTH HAZARDS ASSOCIATED WITH HOT WEATHER WORK

Heat exhaustion and **heat prostration** are different names for the same illness. They are caused by a victim failing to drink enough water to replace fluids lost to sweat when working in a hot environment. Symptoms include the following: cold, clammy skin; fatigue; nausea; headache; giddiness; and low, concentrated urine output. Treatment requires moving the victim to a cool area for rest and replacing fluids. **Heat cramps** are muscle cramps during or after work in a hot environment. They occur because of excess body salts lost during sweating. Treatment involves replacing body salts by drinking fluids such as sports drinks. **Heat fatigue** occurs in people who aren't used to working in a hot environment. Symptoms include reduced performance at tasks requiring vigilance or mental acuity. Victims need time to acclimate to the hot environment and training on ways to work safely in a hot environment.

A **heat illness** is any illness primarily caused by prolonged exposure to heat. **Heat stroke** occurs when a person's thermal regulatory system fails. Symptoms include lack of sweating, hot and dry skin, fever, and mental confusion. Victims need to be cooled immediately or loss of consciousness, convulsion, coma, or even death can result. **Sunstroke** is a type of heat stroke caused by too much sun exposure. **Heat hyperpyrexia** is a mild form of heat stroke with lesser symptoms. **Heat syncope** affects individuals who aren't used to a hot environment and who have been standing for a long time. The victim faints because blood flows more to the arms and legs and less to the brain. The victim needs to lie down in a cool area. **Heat rash** is also called prickly heat. It occurs when sweat glands become plugged, leading to inflammation and prickly blisters on the skin. Treatment can include cold compresses, cool showers, cooling lotion, steroid creams, and ointments containing hydrocortisone. During treatment, victims must keep their skin dry and avoid heat.

HEATSTROKE AND APPROPRIATE FIRST-AID MEASURES

Heatstroke is an affliction often occurring in construction zones. Workers must be able to recognize its symptoms and to deliver an appropriate first-aid response. The human body afflicted with heatstroke loses its ability to adapt to heat stresses and to regulate itself appropriately.

- Heatstroke is characterized by a markedly elevated body temperature — generally greater than 104 F.
- A loss of mental alertness and disorientation are likely to occur.
- Hot and dry skin in some instances, though sweating may be profuse.
- Headache, rapid heartbeat, rapid and shallow breathing, and blood pressure vacillation.

Immediate treatment for heatstroke:

- Move the person out of the sun and into a shady area.
- Direct air onto the person with a fan or newspaper. It is necessary to cool the person and lower the body temperature. Cool, damp cloths or sheets may be used, or cool water sprays.
- Attend the heat stroke victim until professional help arrives. Do not leave the heat stroke victim unattended. Unless instructed by a physician, a heat stroke victim should never be sent home alone.

Conducting a Worksite Investigation of Heat Stress

Safety officials conducting an investigation into a worksite where heat stress injuries are reported should take the following steps:

- Review the onsite OSHA Log or any other existing reports of heat stress injury.
- Interview employees to determine if the employer has taken action to protect workers from heat stress. Are water supplies adequate? Are cooling areas available? Does the employer provide training in heat stress indicators?
- Make a visual inspection of the site by walking around it to assess potential problems. Determine the location of heat sources like machinery, furnaces, or boilers.
- Take heat measurements at various locations. Use the heat stress index to determine the WBGT.

Measuring for Heat Stress Response

Individual heat stress monitoring is recommended when the worker is under heavy loads (burning 500 kcal./hr) and the temperature is at least 69.8 degrees F. Oral temperature and water loss must be measured, along with heart rate monitoring.

- Oral temperature must be taken before the employee drinks water. Drinking water before taking the temperature will have the effect of "cheating" the thermometer. The work cycle should be shortened for any worker whose oral temperature exceeds 37.6 degrees Centigrade or 99.68 Fahrenheit degrees.
- Body water loss measurements are taken by weighing the worker at the beginning and end of each day. Body water weight loss should not be more than 1.5 percent of total body weight per day. Fluid intake should be increased if loss approaches that amount.

Controls for Reducing and Eliminating Heat Stress and Thermal Injuries

The keys to reducing heat stress and thermal injuries are stated below:

- Control the source by keeping heat sources away from occupied areas.
- Modify the environment through ventilation, shielding, barriers, and air conditioning.
- Adjust activities by making the work easier, limiting time spent in hot environments, and requiring periodic rest breaks.
- Provide protective equipment such as water-cooled and air-cooled clothing, reflective clothing, protective eyewear, gloves, and insulated materials.
- Incorporate physiological and medical examinations and monitoring to identify high-risk people.
- Develop a training program to help workers acclimatize to hot environments and learn safe work habits.

Controls for reducing and eliminating cold stress and thermal injuries

A cold environment can be measured according to the air temperature, humidity, mean radiant temperature of surrounding surfaces, air speed, and core body temperature of people in extremely cold temperatures. The keys to preventing injury from **cold environments** are as follows:

- Modifying the environment by providing heat sources and using screens or enclosures to reduce wind speed.
- Adjusting activities to minimize time in cold areas and requiring regular breaks in a warm area.
- Providing protective clothing with insulated layers that both wick away moisture and provide a windscreen.
- Providing gloves, hats, wicking socks, and insulated boots to protect vulnerable extremities.
- Allowing employees time to become acclimated to the cold environment.
- Training employees on practices and procedures for staying safe in a cold environment.

HEALTH HAZARDS ASSOCIATED WITH COLD WEATHER WORK

Trenchfoot occurs when a person spends an extended time inactive with moist skin, at temperatures that are cold but not freezing. Bloods vessels in the feet and legs constrict, causing numbness, a pale appearance, swelling, and, eventually, pain. Treatment involves soaking the feet in warm water. However, the numbness can last for several weeks even after the feet are warmed. Chilblains are an itching and reddening of the skin caused by exposure to the cold. Fingers, toes, and ears are the most susceptible. Gentle warming and treatment with calamine lotion or witch hazel can lessen chilblains. Itchy red hives can occur in some people when their bodies develop an allergic reaction to the cold. The hives may be accompanied by vomiting, rapid heart rate, and swollen nasal passages. Cold compresses, cool showers, and antihistamines can help relieve the symptoms.

Frostbite and hypothermia are the most dangerous cold hazards. Frostbite occurs when the temperature of body tissue goes below the freezing point. It leads to tissue damage. The amount of damage depends on how deeply the tissue is frozen. Severe frostbite can lead to the victim losing a damaged finger or toe. Frostbitten skin is usually white or gray and the victim may or may not feel pain. To treat frostbite, the damaged body part must be submerged in room-temperature water so it can warm up slowly. Hypothermia occurs when a victim's body temperature drops below normal. Symptoms include shivering, numbness, disorientation, amnesia, and poor judgment. Eventually, unconsciousness, muscular rigidity, heart failure, and even death can result. Warm liquids and moderate movement can help warm a victim who is still conscious. An unconscious victim needs to be wrapped warmly and taken for medical treatment.

COMMON HAZARDS AND CONTROLS ASSOCIATED WITH VIBRATION AND IMPACT EXPOSURES

Exposure to vibration can be either whole-body or isolated to a body part, most commonly **hand and arm vibration** caused by use of power tools. Vibration jostles nerves and muscles in unnatural ways and, over time, can cause adverse health effects. Repeated exposure to vibration over long periods of time results in **musculoskeletal disorders** such as tenosynovitis and carpal tunnel syndrome. These disorders are the result of inflammation of the tendon and its sheath due to repeated use or vibrational exposure. Whole body vibration exposure can cause more generalized symptoms such as fatigue, headache, and loss of balance. Over time, the effects can be debilitating, yet difficult to diagnose and ascribe to only one task or activity.

Measurement of Vibration

Vibration is measured in meters per second squared (m/s^2) as a measure of acceleration. The higher the value, the greater the exposure to vibration. The harm caused is a **dose-response relationship** in that people exposed to greater vibration for longer amounts of time potentially suffer more effects. Instruments can measure this acceleration for **impact** (a jackhammer, for example) or **non-impact tools** (a saw or other power tool). Manufacturers of power tools publish data on the vibration magnitude of their tools. There are no currently established exposure limits for vibration, so it is important to evaluate vibration exposure as part of a **comprehensive risk assessment**, since actions can be taken to control exposures before harm is caused.

Design of Tools to Reduce Vibration Hazards

The best solution to a safety hazard is always to **engineer** the hazard out of the job. In the case of a job that exposes a worker to vibration through hand tools, first consider how the job could be done with fixed tools. When this is not possible, aspects of tools can be **redesigned** to reduce vibrational hazard. For example, a different tool with lower vibrational risk could be used. Perhaps wheels or disks on grinding tools could be replaced or better balanced to reduce vibration. If the tool is powered by compressed air, perhaps fewer pounds per square inch could be used. This will reduce the vibrational force of the tool. Also check for different hand tools with a lower intrinsic vibration. Inquire with manufacturer representatives to find a better solution.

Vibration Hazard Training

Workers need to be aware of the risks posed by vibrational hazards. They need to be trained in the proper **grip** to use with a hand or power tool. They need to know how to keep their tools in top **operating shape** so less force will be needed for use (for example, making sure blades are sharpened and grinding wheels regularly replaced). Workers should **dress** warmly enough to ensure circulation to their hands or other body parts affected by vibration. Workers should **grip** tools just as hard as necessary; gripping too hard will result in more vibrational energy being transferred to the worker's hands and arms. Workers should wear the recommended **gloves** for the job; some gloves can reduce the impact of vibrational exposure.

Recommended Exposure Limits for Vibration Exposure

There are no currently established OSHA exposure limits for vibration. However, the American Conference of Governmental Industrial Hygienists (ACGIH) has developed a **Threshold Limit Value (TLV)** for hand-arm vibration that recommends a maximum eight-hour day vibration exposure total value of **4 meters/sec^2**. Maintaining this level will keep most workers from developing symptoms associated with vibration exposure. As a point of comparison, jackhammers are in the 8–15 m/sec^2 range, a chainsaw is typically in the 6 m/sec^2 range, and a random orbital sander is in the 7–10 m/sec^2 range. In order to use these tools without causing harm to the workers, job rotation and exposure limitation must be used to reduce exposure to stay below recommendations.

The Globally Harmonized System of Classification and Labeling of Chemicals (GHS)

Hazard communication refers to the Occupational Safety and Health Administration (OSHA) Hazard Communication Standard found in 29 CFR 1910.1200. The Hazard Communication Standard governs the requirements to notify workers of chemical hazards faced at work and to provide information on protection from hazards. GHS refers to an international standard developed by the United Nations to guide hazardous chemical labeling, warning systems, and safety data sheets. The GHS was developed to standardize hazard warning terminology, pictograms, and safety data sheets worldwide so that international commerce could be improved and language barriers overcome. The

OSHA Hazard Communication Standard has recently been updated to include the requirements of the GHS. As of 2015, all workplaces in the United States are required to have safety data sheets available on site that conform to the GHS system and to use these in their notification and training programs.

SDS (Safety Data Sheets)

Safety Data Sheets (formerly Material Safety Data Sheets) provide information on the physical and chemical properties of a substance as well as potential health and environmental concerns. OSHA requires that all chemicals be labeled appropriately and that SDS be readily available in the workplace. The hazard communication standard also requires employees to be trained, and for the employer to maintain records of the training given. The format for SDS includes sixteen sections. The required sections are as follows:

- I -- Identification
- II -- Hazard Identification
- III -- Composition/Information on Ingredients
- IV -- First Aid Measures
- V -- Firefighting Measures
- VI -- Accidental Release Measures
- VII -- Handling/Storage Requirements
- VIII -- Physical/Chemical Properties
- IX -- Exposure Controls/Personal Protection
- X -- Stability/Reactivity
- XI -- Toxicological Information
- XII -- Ecological Information
- XIII -- Disposal Considerations
- XIV -- Transportation Information
- XV -- Regulatory Information
- XVI -- Other Information

SDS provide a number of indicators for possible health threats of a particular chemical. They are required to provide all known information regarding carcinogenicity of a substance (known or potential cancer-causing risks). Carcinogenic risks are published in the National Toxicology Program report (NTP), the International Agency for Research on Cancer (IACR), and Occupation Safety and Health Administration (OSHA). Toxicity levels are indicated by numbers called the LD_{50} and the LC_{50}. LD_{50} refers to the dose at which 50% of the test subjects were killed. LC_{50} is the lethal concentration at which 50% of test subjects were killed. Dosages are typically normalized to include the mass of the possible toxin divided by the mass of the test subject. LD_{50} values may also include descriptors that indicate the mode of administration of the dose (intravenously or orally) and the timeframe for death after administration. Limits for exposure to a particular chemical are also provided. These can be measured as the OSHA permissible exposure limit (PEL) and/or the Threshold Limit Values (TLV), which are published by the American Conference of Governmental Industrial Hygienists (ACGIH).

SDS often recommend the usage of chemical protective clothing (CPC). Protective eye goggles with splash guards and air vents should be used when handling chemicals. Face shields should be used when working with large quantities of a substance and are most effective when used in conjunction with safety goggles. If the mode of possible hazard is through contact and/or absorption on skin, appropriate gloves should be worn. Gloves are chosen based upon their permeability to and reactivity with the chemical in use. Personal respiratory equipment may be indicated if fume hoods do not provide adequate ventilation of fumes or airborne particulates. Body protection depends on the level of protection needed and ranges from rubberized aprons to full suits that are evaluated for their permeability and leak protection. Closed-toed protective shoes should always be used when working with chemicals.

REQUIREMENTS FOR LABELS

The term "label" under the GHS of Classification and Labeling of Chemicals refers to the label on the container. Under GHS, it's required to contain certain elements; these requirements apply whether the label is affixed by the manufacturer or whether the chemical is placed into a smaller, secondary container in the workplace. The label must include the identification of the chemical, the manufacturer's name and contact information, the applicable GHS pictograms, the applicable signal words (either "danger" or "warning," as applicable), and precautionary statements (measures to reduce risk from exposure to the chemical).

PICTOGRAMS

The pictograms used in the GHS system are simple pictures used to convey hazards posed by the chemical. They are meant to be universally understandable by people with diverse language and reading fluencies. They are as follows:

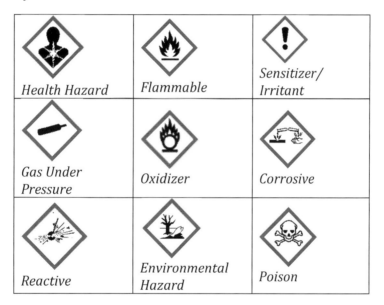

SIGNAL WORDS

Under the GHS hazard communication and safety data sheet system, the term "signal word" is used to describe one word that summarizes the degree of danger posed by the substances. There are only two signal words: "danger" and "warning." The word "danger" is used for more hazardous substances that present immediate hazards such as flammability, reactivity, poison, and so on. The word "warning" is used for lesser hazards such as irritants, environmental hazards, and less toxic substances. The signal word is used on the label to provide a quick and easily understandable indication of the degree of hazard posed by the substance.

BASIC SAFETY THROUGH DESIGN

The hierarchy of controls refers to the preferred methods of controlling health and safety hazards. In order of most preferred first, these include the following:

1. Elimination is completely eliminating the hazard through process changes.
2. Substitution is substituting a lesser hazard, for example, changing from an organic solvent parts washer to an aqueous parts washer.
3. Engineering controls are physical modifications to a work station that serve to reduce or eliminate hazards. An example of engineering controls is to install duct work and exhaust ventilation to remove fumes from the breathing zone of the worker.
4. Administrative control refers to worker management as a means of controlling the hazards. An example of administrative controls is job rotation to limit an employee's exposure to repetitive motion.
5. PPE is garments or auxiliary equipment used to protect workers. Examples of PPE include respirators, gloves, Tyvek suits, welding hoods, and steel-toed boots. This is the last item in the hierarchy because PPE is often uncomfortable to wear, and its effectiveness relies on employee compliance, whereas the other methods of hazard elimination do not.

ELIMINATION OF HAZARDS AND SUBSTITUTION TO MITIGATE HAZARDS

Elimination of hazards refers to making process changes that completely eliminate the hazard rather than developing a work-around or protective device. An example would be changing a process to eliminate solvent use. Another example of elimination of hazards would be to change the way parts are delivered to an assembly line that eliminates the need to lifting heavy boxes. Substitution is another mechanism to mitigate hazards. Examples of substitution include substituting chemicals in a process that are of lower toxicity than the original chemical or substituting a tool that requires awkward, forceful grasping with one that uses a more ergonomically favorable grip.

ENGINEERING CONTROLS

Engineering controls to mitigate occupational hazards are changes in the way the process is designed or the physical controls on the process that make it unlikely or impossible to be injured by the hazard. An example of engineering controls is installing ventilation systems to remove hazardous vapors from the worker's breathing zone, eliminating chemical or hazardous dust exposure. Another example is to install permanent access platforms with proper guardrails to provide elevated machinery access that will eliminate the hazard of using an aerial lift or ladder to access at heights. Another example of an engineering control is to install a switch on an electrical testing device or press brake that requires two hands to activate it; this eliminates the possibility of inserting one's hands into the point of operation while it is operational.

ADMINISTRATIVE CONTROLS

Engineering controls for hazards require a redesign of a work process or physical environment that makes it very unlikely that employees will continue to be exposed to the hazard. Examples of engineering controls include installation of ventilation systems to remove hazardous dust from the workplace, installing assistive lifting devices in a manufacturing assembly line to eliminate lifting, and reconfiguring an electrical testing station so that two hands are needed to activate the on switch, therefore, keeping the worker's hands out of electrical shock hazards.

PHYSICAL AND WORK PRACTICE CONTROLS

Physical controls are measures that alter the physical work environment with the goal of reducing or eliminating risk. Examples of **physical controls** include providing machine guarding to protect from pinch points and gears, barriers to prevent entry into confined spaces, installation of ventilation hoods that remove contaminated air from the workplace, and two-finger switches that prevent a machine operator from placing his or her hands in the zone of operation. **Work practice controls** are administrative in nature and seek to control risk by policies and procedures rather than physical barriers. Examples of work practice controls are as follows: implementing a policy that any lifting of objects more than fifty pounds requires two people or instructing an employee operating a grinder to always wear a face shield during the grinding. Clearly, work practice controls are potentially not as effective as the physical controls as they rely on employees following instructions and policies, whereas physical controls do not present the option to circumvent the control and be exposed to the risk.

RISKS ASSOCIATED WITH MULTIPLE TRADES WORKING SIMULTANEOUSLY IN WORK AREA

PROJECT PLANNING TOOLS

Project planning involves creating a written plan of the steps to complete a project with a timeline. There are several software packages and visual representation tools available for project planning. These tools can be used to lay out a **project timeline** step by step, allowing potential hazards and risks posed by simultaneous work to be assessed. For example, a **Gantt chart** represents various tasks across time with bars showing how long each task lasts, enabling visualization of tasks that occur at the same time. Then one can list hazards posed by each activity during a given time period, linking to the areas the work activities will be occurring.

BYSTANDER HAZARDS

When multiple tradespeople are working on a site, communication is essential to avoid bystander hazards. When a person not directly involved in the work activity is inadvertently injured by the work activity, this is a **bystander hazard**. For example, a bystander who is unaware that heavy equipment is working in the area could walk into the equipment path and be struck. A bystander could also be struck by flying debris during a demolition activity or by falling items from overhead work. Measures to prevent bystander hazards include holding a central **site safety and communication meeting** each day to ensure that all workers are aware of the day's planned work, discussing potential hazards, and planning controls.

PRINCIPLES OF ERGONOMICS AS APPLIED TO CONSTRUCTION PRACTICES AND MATERIAL HANDLING

Ergonomics is the study of the use of the human body in the work environment. It refers to how the body is configured, the motions it is required to take, and the forces applied to muscles and bones by the work. Ergonomics seeks to adapt the work demands to make them as easy as possible for the body to endure to prevent injury. Anyone who has participated in physical activities and been sore the next day understands the effects of overexertion. Overexertion at work over long periods of time can lead to injuries that can be costly to treat. Even in the absence of an injury situation, improved ergonomics can improve productivity and comfort in the workplace.

Good **ergonomics** improves output and performance and reduces error and accidents by making the environment comfortable and user-friendly. The four general principles of ergonomics that apply to safety engineering are as follows:

- **People versus machines**. This principle states that people are better at some jobs, such as reasoning inductively and handling unexpected occurrences, while machines are better at other jobs, such as repetitive operations and deductive reasoning.
- **Change the job, not the person**. People have limits, and not recognizing those limits can cause errors, hazards, and accidents. It is better to change a job, equipment, or environment to fit the person rather than trying to change the person to fit the job.
- **Work smart**. Productivity can be improved and errors reduce by finding new and better ways to do a job.
- **People are different**. People differ in their age, height, weight, reaction time, strength, coordination, attitudes, etc. Designers and managers need to adjust jobs accordingly.

ORGANIZATIONAL, BEHAVIORAL, AND PSYCHOLOGICAL INFLUENCES ON ERGONOMIC INJURIES

Organizational, behavioral, and psychological factors can influence the likelihood and severity of ergonomic injuries. **Organizational factors** include how the work processes are designed and laid out, the time demands that are placed upon the workers, and how long the cycle times are. Workstation design is the most important organizational factor. **Behavioral factors** that impact ergonomic injuries include observing proper lifting behavior, observing the buddy system for lifts over 25 pounds, and observing proper neutral postures. Poor postures may cause the employee to feel pain, which results in uneven muscle strains. The most important **psychological contributor** to ergonomic injuries is stress. Stress and perceived lack of control lead to poor job satisfaction and promote a feeling of monotony at work. These stress factors can increase perceptions of pain, decrease blood flow, create muscle tension, and therefore increase the likelihood of an ergonomic injury.

THE FOUR MAJOR ERGONOMIC STRESSORS

When considering the likelihood of an ergonomic injury, tasks must be evaluated with regard to four major stressors.

- **Repetition**—how often does a certain movement have to be repeated during the work day? Higher numbers of repetitions create more stress on bones, muscles, and tendons, thus increasing likelihood of injury.
- **Force**—how much force must be applied when performing the task? Force also includes how much weight must be lifted. Any lifting greater than 25 pounds is considered heavy. The heavier the lifting (or force), the greater the likelihood of injury.
- **Posture**—this refers to the orientation of the body or body part. Neutral postures are the least stressful. The farther from neutral the body is required to be, the greater the likelihood of injury.
- **Vibration**—this refers to the amount of vibratory movement required. Vibration is generally caused by holding power tools (e.g., a drill or jackhammer). Exposure to vibration for long periods of time can damage nerves.

All of these **risk factors** combine to produce an overall risk of injury. The common theme among all of the risk factors is the degree to which the body is required to *deviate* from its natural, neutral state. The greater the deviation or demand on the body, the greater the likelihood of injury. An additional contributing factor is the amount of rest allowed between exposure to risk factors. Over time, with insufficient rest, tendons and ligaments can become inflamed and painful and can lead to an ergonomic injury. Ergonomic injuries sometimes require surgery to correct. This is a cost to both the employer and the employee.

Proper Lifting Technique

Use of proper lifting techniques can prevent back strains in the workplace. The goal of proper lifting is to put as little stress on the spine as possible. The spine is meant to be straight, and lifting with a bent, curved-over spine puts extreme stress on it, even if the weight lifted is relatively small. To properly lift, place the feet hip-distance apart, and face the object to be lifted. Squat to pick up the object, then pull the object close to the body at the waist to move it. Stand up, using the strong thigh muscles to accomplish the work of lifting. Do not twist the spine when lifting or moving the object. Reverse the process to put down the object. Peer assessments should be conducted regularly to reinforce proper lifting techniques as people tend to get lazy and revert to poor lifting techniques.

Neutral Standing Posture

A neutral standing posture can help prevent fatigue and muscle strain. The feet should be kept shoulder-width apart for optimal support. Shifting one's weight from one foot to the other is a sign of fatigue that indicates inattention to neutral standing posture. Hips should be in line with the shoulders and the knees. The arms should hang comfortably at one's side, and the shoulders should not be hunched or tensed. The neck should stay straight (head not bent over), and the spine should stay straight, not twisted.

Repetitive Stress Injuries

A repetitive stress injury is caused by using the same muscles, tendons, and bones repetitively several times over the course of a work day. Repetitive stress injuries occur because of strains on muscles, tendons, and ligaments. If caught early, and the stress is eliminated early, they are almost always reversible. However, repetitive stress injuries can also lead to employees having to have surgery and spend weeks away from work, followed by months of light duty. A common example of a repetitive stress injury is carpal tunnel syndrome, which is a disorder of the tendons and nerves in the wrists and hands experienced by some workers who must do repetitive work with their hands (such as data entry operators, typists, or hand assembly workers).

TECHNIQUES TO DECREASE LIKELIHOOD OF INJURY: Repetitive stress injuries are caused by using the same motion (and thus, the same muscle groups and bones) over and over again. Another contributing factor is the force applied to the motion. To reduce the likelihood, it is important to quantify the loads and the numbers of repetitions. Configure the work areas to ensure that items used most often are within close reach and that long reaches or limb extensions are only required infrequently. Break up loads into smaller weights. It is also important to evaluate whether employees' work stations are configured to ensure neutral postures are used most of the time. In addition, job rotation can also be important to changing the motions employees have to make over the course of the work day.

Minimizing or Eliminating Lifting Hazards

Minimizing or eliminating lifting hazards should be pursued as an effective injury prevention tool. Tools to do this include ensuring that all loads lifted with regularity are less than 25 pounds, and heavier loads that must be lifted must use a buddy system (two-person lift) or assistive device (forklift, drum lifter, or vacuum-powered lifter). Pallet jacks that can be height adjusted to the work level can minimize or eliminate the need to carry a package from floor height to work table height, which decreases the risks posed. Carts can be used to transport items that must be moved from one location to another rather than carrying them.

NIOSH LIFTING EQUATION

The NIOSH lifting equation uses several factors to calculate a **recommended weight limit (RWL)** for a certain task. This RWL is the maximum weight that almost any worker can lift over an eight-hour shift without increased risk of musculoskeletal injury. The equation is as follows:

$$\text{RWL} = (51 \text{ lbs.}) \times (\text{Horizontal Multiplier}) \times (\text{Vertical Multiplier}) \times (\text{Distance Moved}) \\ \times (\text{Asymmetry Angle}) \times (\text{Frequency Multiplier}) \times (\text{Coupling Multiplier})$$

The horizontal and vertical multipliers are how far the load must be moved in the horizontal and vertical directions. The asymmetry angle is the number of degrees from vertical the body must be moved. The frequency multiplier ranges from 0.2 to 15 times per hour. The coupling multiplier is a measure of how well the person's grip is on the load to be moved, ranging from 1 (good) to 3 (poor).

RAPID UPPER LIMB ASSESSMENT

A Rapid Upper Limb Assessment (**RULA**) is an ergonomic evaluation method that can be used to screen and quantify the degree of ergonomic risk from a given task or work flow. The tool requires observation and evaluation of a work task with regard to body posture, required application of force, and degree of repetition in the tasks. Body postures receive higher scores the more they deviate from neutral. Force or lifting is evaluated from 1 kilogram to 10 kilograms; any forces or lifting greater than 10 kilograms is assigned the highest score. Repetition is scored from intermittent to frequent. Each body part evaluated is assigned a number from 1 to 7. Scores of 1 to 2 are considered acceptable ergonomically, 3 to 4 may require change, 5 to 6 require change in the near time frame, and scores of 7 require immediate evaluation and change. The RULA is most appropriate for seated or sedentary tasks that do not involve the entire body.

RAPID WHOLE-BODY ASSESSMENT

A Rapid Whole-Body Assessment (**RWBA**, also referred to as Rapid Entire Body Assessment or REBA) is an ergonomic evaluation tool used to assess degree of ergonomic strain or risk to the entire body from a given work task sequence. It is used as a screening tool to provide a relative risk ranking that can be used as a guide to determine which work tasks to redesign first. Assessment is similar to that of the RULA, and is expanded to the entire body—neck, arms, legs, trunk, and wrists. Scores obtained in the RWBA range from 1 to 15 and are evaluated as follows:

- 1–3: low risk
- 4–7: medium risk, action may be necessary in the future
- 8–10: high risk, action is necessary in the short term
- 11–15: very high risk, action is necessary immediately to redesign the work task

The RWBA is best applied to tasks that involve the *entire body* rather than just the upper body.

REQUIREMENTS, USAGE, AND LIMITATIONS OF PERSONAL PROTECTIVE EQUIPMENT

Personal protective equipment is any specialized clothing or equipment worn or used to protect a person from hazards. Personal protective equipment can include HAZMAT suits, goggles, gloves, respiration equipment, hard hats, and more. Personal protective equipment can form an essential part of a safety plan, but it should never be a primary means for controlling hazards. Personal protective equipment forms a barrier between the user and a hazard, but it does not remove the hazard. It is far better to remove the hazard, if possible. Personal protective equipment has limited success because the user must have the right equipment and know how to use it properly, often in an emergency situation. In addition, the personal protective equipment must fit properly and must be well-maintained.

In order to be effective, a program for personal protective equipment must include detailed, written procedures that address how to select, manage, use, and maintain personal protective equipment. These procedures must be enforced and supported by management and should include standards and rules for the following:

- Wearing and using personal protective equipment.
- Inspecting and testing personal protective equipment to ensure it is in good condition and working properly.
- Maintaining, repairing, cleaning, and replacing personal protective equipment.

Another important element of a personal protective program is to ensure that users understand and accept the importance of personal protective equipment. Allowing users to participate in selecting the personal protective equipment they will wear can help them "buy into" the program and be more likely to use their personal protective equipment when it is needed.

Four areas where personal protective equipment is a major factor in accident prevention are in the areas of head protection, face and eye protection, feet protection, and hand protection.

- Hand Protection: Gloves should be appropriate to the task. To prevent concrete poisoning, non-porous rubber gloves must be used. Specific glove designs should be used for welding, whereas another non-conductive type is used for electrical work.
- Head Protection: Hard hats are the rule and should be regularly inspected for damage or wear.
- Eye and Face Protection: A combination of safety glasses and facial shields may be used for construction site activities like electrical work, welding, cutting, grinding, and nailing.
- Foot Protection: Steel-toed shoes and boots are the rule when working around forklifts or around other heavy and rolling objects. Safety plated shoes can prevent sharp objects from penetrating the soles, and non-slip soles provide better grip on walking surfaces.

PPE FOR HANDS, FINGERS, ARMS, FEET, AND LEGS

Hands, feet, arms, fingers, and legs need protection from heat and cold, sharp objects, falling objects, chemicals, radiation, and electricity. Gloves and mittens protect the hands and fingers and can even extend up the wrist and arm. They can be made of different materials according to the protection needed: for example, lead is used for radiation protection and leather for protection from sparks. Gloves can be fingerless to protect just the hands; finger guards are used to protect just the fingers. Creams and lotions also protect hands and fingers from water, solvents, and irritants. Feet can be protected by the appropriate type of shoes: safety shoes with steel toes and insoles; metatarsal or instep guards; insulated shoes; and slip-resistant, conductive, or non-conductive soles. Rubber or plastic boots provide protection against water, mud, and chemicals while non-sparking and non-conductive shoes are useful for people working around electricity or where there is a danger of explosion. Shin guards and leggings protect legs from falling and moving objects and from cuts from saws and other equipment.

Work gloves

Work gloves fall into four basic categories based on the type of materials used in making the glove and the type of materials they are designed to handle.

- Gloves made of leather, canvas, or metal meshes are designed for general-purpose material handling. Most will protect from splinters or from sharp edges. Metal mesh gloves can protect from puncture but should never be used near electrical lines.
- Coated Fabric gloves are often used for work like brick-laying. They may offer some protection from slippage and from chemicals but are generally considered "light duty" and not protective in extreme situations. Coated gloves are made from fabric and an applied coating of rubber or vinyl.
- Insulating Rubber Gloves are specifically designed to protect against electric shock. Most insulating gloves have the limitation of not being able to protect against other hazards like punctures. In most cases, rubber insulating gloves are covered with a second pair of durable leather gloves which can protect from pinching and punctures or similar work problems
- Chemical and Liquid-Resistant Gloves are made according to the materials they are designed to handle. Design of these gloves may vary. Some may cover the hands, wrists, and beyond the elbows.

PPE for the head and face

Personal protective equipment for the head can protect wearers from being hit by falling or flying objects, from bumping their heads, and from having their hair caught in a machine or set on fire. Helmets bump caps, and hard hats are examples of this type of head protection. Hoods and soft caps are another type of head protection that also protects the face and neck. Hoods may include hardhat sections as well as air supply lines, visors, and other protective features. They provide protection from heat, sparks, flames, chemicals, molten metals, dust, and chemicals. Head protection can also aid sanitation by keeping hair and skin particles from contaminating the work. This is especially important in processes involving food and clean room work. Hairnets and caps offer this type of protection. Face shields and welding helmets protect the face from sparks, molten metal, and liquid splashes. They should always be used in conjunction with eye protection such as goggles or spectacles.

Hard hats and helmets

There are two helmet types specifically designed to protect the wearer from electric shock: Class E and Class G. A third type, Class C, may be worn by utility workers but offers no protection against electrical shock; its primary purpose is to protect from head injury due to bumping or dropped objects.

- Class E: This type of helmet offers the greatest protection against electrical shock from high voltage conductors. Class E (think E, for electrical) helmets are proof tested at 20,000 volts. Class E Helmets are tested for being able to sustain an amount of force, their ability to repel electrical shock without leakage, and the capacity to function without burn-through.
- Class G: This type of helmet will protect against electrical shock but not of a voltage exceeding 2,200 volts, applied for 60 seconds. Class G helmets were formerly known as Class A helmets until the more recent change. Think "G", for general use.
- Class C Helmets afford very little or no electrical shock protection. They protect the head only from bumps, jolts, and dropped objects.

OSHA standards do not legally require the use of hard hats in areas where workers are not exposed to hazards from falling or flying objects or electrical shocks. However, employers may mandate use of hard hats though such mandates may require negotiation with union representatives.

PPE FOR THE EYES

The CHST should be familiar with the various types of eye protection which may be used to protect against specific hazards peculiar to the type of task performed. Eyes need to be protected from flying particles and objects, splashing liquids, excessive light, and radiation. Types of eye protection include spectacles, goggles, and welding helmets.

- **Safety spectacles:** These simple protective eyeglasses are constructed of metal or plastic with impact-resistant lenses. Side shields are available on some models.
- **Goggles:** These are tight-fitting eye protection devices which are designed to completely cover the eyes. Goggles provide protection from impact of flying objects, from dust and splashed chemicals. Goggles may be fitted over personal eyeglasses of the employee.
- **Full-face Welding Shields:** Vulcanized fiber or fiberglass face shields are fitted with a filtered lens. Designed to protect against intense radiant light, they also protect from flying debris producing during the welding, soldering, or brazing process.

RESPIRATORY PROTECTION

Respiratory protection programs are required of employers within any site where airborne materials (including gases) may pose a threat to worker health. When a worker enters the respiratory program of an employer, the worker is subject to a medical evaluation (by a licensed health care practitioner or LHCP) of his fitness to use and wear respirators. Respirators are designed for different purposes:

- An atmosphere-supplying respirator refers to a means of supplying the user with an independent air supply usually contained in a tank or storage device. These types are designed for areas when it is impractical or dangerous to use ambient air.
- An air-purifying respirator is one that uses ambient air and filters it through a canister or cartridge to remove toxic elements. This type of respirator is useful when air contamination is not of such volume and extent as to overwhelm the filtering cartridge.

The chief concern in choosing the appropriate respirator is the material hazards presented by the worksite. Air purifying respirators cannot protect against gases, in other words. Other concerns may also impact the CHST's decision in choosing a respirator:

- The physical configuration of the jobsite is a factor in choosing the correct respirator. Tight areas and close quarters may not permit the use of SCBA due to their size and bulk. Similarly, the type of respirator with a long hose is not suitable for working around machinery or other obstacles.
- The workers medical condition is another factor determining choice of respirator. Negative pressure respirators wouldn't be appropriate for workers who have asthma, for example.

Worker comfort should be considered also. A comfortable worker is a more productive worker. Masks and helmets should meet fit requirements.

BODY PROTECTION

Personal protective equipment for the body provides protection from hazardous materials, biohazards, heat, fire, sparks, molten metal, and dangerous liquids. This clothing is often combined with other types of personal protective equipment, such as respiratory equipment and eye protection. Types of personal protective equipment for the body include the following:

- Coats and smocks to protect clothing from spills
- Coveralls, which may include hoods and boots
- Aprons, which protect the front of a person from spills and splatters
- Full body suits for working with substances that present a danger to life or health (may include cooling units to help lower the wearer's body temperature)
- Fire entry suits
- Rainwear
- High-visibility clothing for people working on road construction or in traffic
- Personal flotation devices for people working around or on water
- Puncture-resistant or cut-resistant clothing for protection from ballistic objects, power saws and other cutting equipment

Some personal protective equipment is designed to be disposable, especially clothing contaminated with hazardous materials.

Storage and maintenance

Proper storage and maintenance of personal protective equipment is important for personal hygiene and for proper operation of the equipment. The following covers storage and maintenance procedures for respirators, hearing protection, and gloves:

Respirators – Respirators must be stored in a sealed plastic bag to protect them from being contaminated by dust. Best practice dictates that respirators are only stored after they are thoroughly cleaned and dry. Maintenance of respirators requires that all inhalation and exhalation valve flaps be in place and replaced if worn, and that filter cartridges be replaced according to a pre-established schedule or whenever breakthrough is detected.

Hearing Protection – Ear plug hearing protection should only be inserted using clean hands. If ear plugs are soiled or fall on the ground, they should be replaced. They should be stored in a plastic bag away from potential dust contamination. Replacement of disposable earplugs should be done weekly if they are worn daily, or if they seem to be losing elasticity. Ear muff hearing protection needs only to be cleaned off periodically. They should be replaced if they are cracked or worn and do not provide a good seal around the ears.

Gloves – should be stored in a locker or other area away from dust. They must be examined before each use to ensure there are no holes or tears. Gloves should be replaced when worn out.

Inspection

Once the proper personal protective equipment is selected in accord with a job hazard analysis, it must be properly inspected and maintained. Whenever possible, employees should be assigned their own PPE devices and should not have to share with another employee. Certain types of PPE must be inspected for proper configuration before each use, such as fall protection equipment and respirators. High-value and high-risk PPE should be inventoried and documented inspections conducted (for example, fall protection equipment and SCBA respirators). Assigning the equipment to a particular person is essential so that any deviations from inspection procedures or misplaced equipment can be identified and the offending employee counseled. All PPE used for emergency purposes must be inspected at least monthly, and these inspections should be documented to show they have been conducted.

BASIC TESTING AND MONITORING EQUIPMENT

INDUSTRIAL HYGIENE SAMPLE AND INDICATOR MEDIA

There are three basic types of sampling methods used to determine concentration levels of gases and vapors: a spot-check or snapshot sample, a long-term sample, and a passive sample. A spot-check is a quick sample of the air in a specific location which is good for determining exposure risk at the moment of the sample. A more thorough evaluation of the air environment for long-term exposure is the long-term sample where a device samples the air over a period of time of eight hours or more to determine the contamination level for an average work day or the length of the sample. The passive sample method is where a worker wears a sample device that simply warns when contamination levels are too high. Although the passive sample seems like the simplest method for keeping workers out of harm's way, these devices provide no measurement capability for preventive measures.

SAMPLING EQUIPMENT

Below are several hand-held and portable instruments that are used in industrial hygiene applications:

Noise meters and dosimeters- are used to conduct sound level exposure surveys, and are worn for an entire shift to determine whether the OSHA Permissible Exposure Limit of 85 decibels was exceeded. Results and reports can be downloaded to a computer for storage and documentation.

Gas meters –are used for real-time analysis of various gas levels, such as, oxygen, carbon monoxide, and hydrogen sulfide. These are mandatory for permit-required confined space entry.

Gas detection tubes – are glass tubes containing a chemical-specific reactive substance used to measure concentrations of chemicals in the atmosphere. They are available for a variety of chemicals used in industry and provide an accurate assessment of exposure. Some are designed to be used with a hand-held pump and others are worn by the employee and sample the air by passive diffusion.

Photoionization detectors – are portable hand-held instruments that measure the concentration of volatile organic compounds in the air. They are not chemical-specific but measure all VOCs in the surrounding air. Photoionization detectors provide a way to assess exposure levels and establish level of respiratory protection needed in a given environment.

COLORIMETRIC INDICATOR TUBES

Colorimetric indicator tubes are a popular method for monitoring for airborne contaminants since the tubes provide a quick response and are easy to use. The tubes work by drawing a certain amount of air through the tube, passing through a layer of cotton and a conditioning filter, then interacting with an indicator ruler which changes color according to the concentration, and then back out through another cotton plug. The color changing indicator line is read against a direct reading scale which displays the concentration usually in parts per million. The colorimetric indicator tubes provide an excellent warning signal for certain hazardous contaminants and can be used for monitoring for several different chemicals, but the calibration of the tubes is a concern as only the manufacturer can do so, making inaccurate readings a higher risk factor. Once used, these tubes must be disposed of as hazardous waste.

PH STRIPS

pH is a scale from 1 – 14 that is a measure of acidity (free hydrogen ions in solution) or alkalinity (free hydroxyl ions in solution). pH strips are plastic or paper strips coated with a substance that changes color based on the amount of hydrogen or hydroxyl ions in a solution. Pure water is considered neutral with a pH of 7. pH strips can be used for quick determination of alkalinity or acidity of a liquid. If the pH is below 2 or above 12 then it is classified as a corrosive hazardous substance. pH strips are used on storm water runoff for monitoring purposes. pH analysis has a fifteen-minute holding time, which is not long enough to get a sample to the laboratory, making the test strip a useful alternative. pH determinations for regulatory compliance should ultimately be confirmed by use of a calibrated pH meter. pH strips must be protected from humidity and other gases to maintain accuracy.

LIGHT METERS

Light is measured in units of lux, which is one lumen per square meter. A lumen is the intensity of one candle in a cone-shaped area. Light levels can be measured with hand-held instruments. Appropriate light levels are important for various tasks in order to reduce eye strain and avoid injury. To get a sense of the scale of lux values, a bright summer day is about 100,000 lux, an overcast summer day is about 1,000 lux, and twilight is about 10 lux. In occupational settings, a normal office environment should be at 250 – 500 lux. A supermarket or mechanical workshop should be approximately 750 lux. Detailed drawing work requires 1,500 – 2,000 lux for comfort. Consideration should also be given to the type of light used, as different types have different qualities.

CONSTANT-FLOW SAMPLING PUMPS

Direct reading instruments take snapshots of the air quality and provide almost instant readings of the concentration levels of specified contaminants in the air. These readings are reliable enough for detection and early warning of possible hazardous exposure levels of the contaminants, but for complete accuracy and regulatory compliance verification, the constant flow sample pump is the measurement device used by OSHA. The constant flow air pump draws a measured amount of air through a filter medium for a given period of time to measure exposure levels for up to eight hours in a single day. The contaminants captured by the filter medium are analyzed by a laboratory to measure the concentration and determine the time-weighted exposure levels. This most accurate measurement of the contaminant level is used to determine whether or not a facility is in compliance with OSHA permissible exposure level limits regulated by OSHA.

Emergency Preparedness and Fire Prevention

PROPER FIRE PROTECTION AND PREVENTION METHODS

CLASSES OF FIRES

The United States fire classification system identifies the first three primary classifications of fire types as A, B, and C.

- Class A: This is the most common type of fire and refers to the ignition of combustible materials like wood, cloth, and paper. Class A fires can be extinguished with water but the use of Class A rated fire extinguishers works as well.
- Class B: This type of fire is the result of ignition of flammable liquids. Water will not extinguish Class B fires. A jet of water may have the opposite effect, forcing the fire to spread or splatter. Class B fires should be put out with Foam, CO2, or Dry Chemicals.
- Class C: Class C fires are electrical. One of the first responses should be to cut off the power source, reducing the fire to Class A type of event.
- Class D: Metals fires can originate with flammable metals like titanium and magnesium. NFPA recommends dry powder to extinguish them. "Dry Powder", it must be remembered, is different in composition from "Dry Chemical". The terms "dry powder" is not to be confused with "dry chemical." Each of these compounds will have a different effect upon a fire.
- Class K: This refers to fires from cooking oils. Do not apply jets of water as this will cause the fire to spread and splatter. Class K fires may be put out with Dry Chemicals from suitable fire extinguishers.

FIRE EXTINGUISHER INSPECTION CRITERIA

Fire extinguishers must be fully **charged** and must have been charged with fire suppression agent within the last year to ensure they are effective. To inspect a fire extinguisher, first note that it is in the correct **location**. Best practice is to mark the locations of fire extinguishers with red arrows that point to the extinguisher. The signs should be mounted high enough on the wall to be spotted from a distance in an emergency. Check that the **pin** is intact and that the **seal** has not been broken. Check the **gauge** that indicates whether the extinguisher is fully charged. Check that there is an **inspection tag** in place, and that the recharge date is within the past year. The recharge date will usually be indicated by punching the month and year of the last recharge. Finally, check the back of the inspection tag to confirm that each **monthly inspection** has been completed and add initials and date for the current inspection.

FIRE SPRINKLER INSPECTION

The National Fire Protection Association publishes **NFPA 25**, the Standard for the Inspection, Testing, and Maintenance of Water-Based Fire Protection Systems. There are separate inspection elements for quarterly, annual, and five-year inspections. Inspections must be performed by a trained and competent person. The quarterly inspection must include operation of the water flow alarm devices, the valve supervisor signal devices, and the integrity of the fire department connections. The sprinkler heads must be inspected for proper installation, evidence of damage or leaks, and proper clearance (1 inch).

COMPONENTS OF EMERGENCY ACTION PLANS

Emergency action plans should be tailored to a specific work site in the construction industry. Plans should include the location of the nearest **hospital** or clinic that is available to treat injured workers, locations of **emergency equipment** such as fire extinguishers and spill kits, and the location of the **assembly area**. Although these are elements common to all emergency plans, it is important to realize that this information will change for each site and should be updated accordingly. In addition, emergency planning should take into account that multiple activities may occur on a site, requiring coordination and communication. Special attention should also be paid to the location of **utilities** and any disruption in utilities that may be caused by the construction activities.

EMERGENCY EVACUATION MAPS

An emergency evacuation map should be created for each work site, and should be posted at visible and accessible areas near exits. An emergency evacuation map should show the basic **outline** of the structure, including the interior walls. It should show all interior and exterior **doors**, and the direction that they open. It should highlight available **exit routes** from each occupied workspace. The map should show the location of the **Assembly Area**. The location of **emergency equipment** should also be marked, such as fire extinguishers, eye washes, emergency spill kits, and safety showers.

SITE RISK ASSESSMENTS

Risk assessment is an essential element of an emergency action plan. Risk assessment is necessary to anticipate what types of emergencies may occur and should be planned for. For example, depending on the geographical location, different **weather emergencies** must be planned for. In some areas of the country, plans must be made for extreme weather events or tornadoes, whereas in other parts, planning for wildfire and earthquake are necessary. Additionally, the risk assessment will help to plan for potential **health and safety emergencies**. For example, if trenching will be conducted, possible emergencies and responses associated with trenching must be assessed. Overhead hazards must be anticipated and planned for. If hazardous chemicals are in use, it is important to assess the potential size of a hazardous materials spill and to have the proper response equipment on site. Many other risk assessment considerations can be made as well. An emergency action plan attempts to answer the question of what to do if something goes wrong.

COMMON ELEMENTS OF RESPONSE PLANS FOR ENVIRONMENTAL HAZARDS

An environmental emergency response plan should include information on how to prepare for and respond to a **chemical spill**. The plan must include a description of where chemicals are stored, what types of chemicals are on site, and where the spill response equipment is located. The contents of each **spill kit** should be included and these must be regularly inspected to ensure the supplies are still intact. The response plan should describe how to clean up simple spills and the notification requirements for spills that may be too large for on-site personnel to clean up. The emergency response plan should be used as a training tool and a reference document but **training and drills** must be conducted to prepare employee responders in advance of an emergency.

WHEN TO NOTIFY AUTHORITIES

If a hazardous substance or chemical is released off-site into a waterway or could potentially contaminate a waterway or sewer system, regulatory authorities must always be notified. Both the US EPA and US DOT define a "**Reportable Quantity**" as the amount of a hazardous substance that, if released into the environment, requires reporting to the National Response Center. Reportable quantities may be found in the **US DOT Hazardous Materials Table, Appendix A (49 CFR**

§172.101). A release to the environment means on land, to the air, or into a waterway. Pure chemicals, chemical mixtures, and EPA hazardous wastes each have specific reportable quantities.

AGENCY CONTACT INFORMATION TO INCLUDE IN ENVIRONMENTAL RESPONSE PLAN

An environmental emergency response plan must include the contact information for the **company emergency coordinator** and **alternative emergency contacts** if that person is unavailable. To act as emergency coordinator, the person must be fully aware of all hazardous materials at the site and have complete site access, including keys to access all locked areas and codes to disarm any alarms in case of emergency. The emergency response plan should also include **agency contact information** for agencies that may need to be notified of a chemical emergency or spill. This includes the sewer agency, the water quality control board for notifications of stormwater pollution, the US EPA National Response Center for notifications of chemical spills, the Department of Fish and Game if wildlife is harmed, and non-emergency contact information for the local fire department.

EMERGENCY RESPONSE SYSTEM

INCIDENT COMMAND SYSTEM (ICS)

An ICS is a standardized, multi-agency, management system designed to apply to all types of hazards from small to catastrophic. The task was mandated by Homeland Security Presidential Directive 5 and is now regulated by the National Incident Management System (NIMS). OSHA rule 1910.120 mandated the implementation of an ICS for all companies that are involved with hazardous materials. FEMA (http://training.fema.gov) provides extensive training materials to help in the development of an ICS. Deployment of an ICS is required when a hazard or natural disaster has a high potential of resulting in harm to humans, facilities, or the environment. Utilization of an ICS helps ensure safety of people and cost-effective use of resources in order to effectively manage a hazard response program.

The Incident Command System is designed to manage emergency situations effectively and efficiently. It is a standardized approach to command, control, and coordination of all types of emergency response. The structure and format of the ICS is part of the National Response System codified in the National Contingency Plan; and is widely used by government agencies when responding to disasters. The ICS is under the direction of an On-Scene Coordinator. This individual acts to coordinate the involvement of various agencies and serves as a clearinghouse for information related to the incident so that those involved coordinate effectively. The ICS is organized with divisions responsible for operations, planning, logistics and finance, with additional support for safety and information gathering and dissemination.

CRISIS MANAGEMENT

The concept of crisis management refers to an organization or company being prepared to handle all aspects of an unexpected event in order to ensure that the situation is properly **managed** and that employees, interested parties, and the public are properly **informed and cared for**. Response to **emergencies** is one type of crisis management that an organization should be prepared for. It is important to anticipate potential events in planning for a response. Consideration must be given to **communication** inside and outside of the organization. Consideration should also be given to the **short-term response** to a crisis (what to do during the emergency) and to the **long-term and follow-up responses** to a crisis (how to rebuild and who to notify).

Business continuity planning can result in a smoother response to crisis management. Although it may not be possible to consider every possible emergency, consideration can be given to fire, earthquakes, tornadoes and other weather events, chemical spills, and disruption in the workforce due to pandemic illness or a labor strike. Envisioning the **potential scenarios** is a good first step to take. Consideration should be given to handling disruptions in internal and external communication, and conducting business if information technology or computer systems are disrupted. **Drills or table top exercises** can be used to simulate a crisis so participants can practice their responses. This type of planning can also help to reveal shortcomings in the planning process and areas of potential improvement.

Emergency First Aid Equipment and Procedures

Eyewashes and Safety Showers

OSHA requirements for emergency eyewashes and showers are found in 29 CFR 1910.151. Requirements for eyewashes and showers are similar. They are required in areas where an employee might be exposed to corrosive or hazardous chemicals, such as strong acids or bases. Eyewashes and showers are required in work areas with anhydrous ammonia and forklift battery charging stations. The path to the eyewash or shower must be unobstructed, and require no more than 10 seconds of travel time from the hazard. Both eyewashes and showers must deliver tepid water. The flow must be able to be sustained for a minimum of fifteen minutes, and the valve that delivers the water must stay open on its own and not shut off unless someone physically shuts it off.

Fire extinguishers

OSHA regulations require a written fire prevention plan that is communicated to all employees (29 CFR 1910.39). Employers with fewer than ten employees may verbally tell their employees the plan. The plan must list the fire hazards present in the workplace, and procedures to control accumulation of flammable and combustible materials. There must be written procedures for maintenance of heat-producing equipment to prevent accidental build-up of heat and ignition of combustible materials. Part of the fire prevention plan must describe where fire extinguishers are located, and what type of fire extinguisher is appropriate for the fire hazard. Fire extinguishers should be located near areas with fire hazards. Employees must receive training in their use. Employees must be trained that fire extinguishers are to be used only for small fires. A back up extinguisher should be available as well. Employees should be instructed to never let a fire back them into a corner; they should always have an exit route when using a fire extinguisher.

AED

An automated external defibrillator (AED) is a device that can be used in an emergency situation when a person's heart has stopped beating (a heart attack). It works by delivering and electrical impulse (shock) that restarts the stopped heart. These should only be used by those that have learned to use them in an approved first aid course. The device consists of two adhesive pads that are wired to the defibrillator. The device can measure the person's heart rate (or lack thereof) and recommend whether the defibrillator needs to be used. These pads are placed on either side of the heart on a dry chest. Press "analyze" to see if the device recommends that an electrical impulse be delivered. If so, be sure not to touch the person, and press the "start" or similar button to deliver the impulse. Then deliver CPR until the emergency responders arrive.

Responding to Media Inquiries

As part of emergency response planning and business continuity planning, it is important to plan **response to media inquiries**. The most important part of the planning is to define who is authorized to speak to the media. Most often, this will be the President or CEO, but could also be an assigned **public relations officer** in a large organization. All other personnel should be instructed to **refer** inquiries to this person. When responding to the media, it is important to be factually accurate without being alarmist. Communication must also be maintained with agency first responders to coordinate responses to media.

POTENTIAL FIRST AID OR MEDICAL NEEDS

AVAILABILITY OF FIRST AID KITS

First aid supplies must be on hand for the trained first aid responders to use when providing first aid. There are no specific lists of supplies that are required by OSHA. The reference for the **minimum contents of a first aid kit** may be found in **ANSI standard Z308.1**, *Minimum Requirements for Workplace First Aid Kits*. This ANSI standard requires adhesive bandages, adhesive tape, topical antibiotic treatment, antiseptic wipes, burn dressing, burn treatment supplies, a breathing barrier to provide CPR, a cold pack, eye covering and eye wash, exam gloves, sterile gauze pads, hand sanitizer, and a triangular bandage that can be used as a sling. There may be specific hazards at a given workplace that require more supplies, or more extensive supplies. For example, if workers are working with glass or ceramics and there is greater risk of cuts, it may be necessary to stock more extensive first aid supplies designed to stop bleeding.

AVAILABILITY OF TRAINED FIRST AID RESPONDERS

The OSHA regulations regarding trained first aid responders may be found in **29 CFR §1910.151**. This standard requires that **trained first aid providers** be available if there is no clinic or hospital in "near proximity." Unless the work site is actually at a hospital or clinic, it is unlikely that a medical facility is near enough to provide the proper first aid when needed. In addition to requiring a person or persons trained in administering first aid, several OSHA regulations require trained personnel in **cardio-pulmonary resuscitation** (CPR). Also be aware that if workers may be providing first aid as part of their job duties, they must be trained in the hazards of **bloodborne pathogens** and be offered Hepatitis B vaccination as is required by the bloodborne pathogen standard.

CPR

CPR should only be performed by trained and qualified people. A well-stocked first aid kit will include a **CPR face shield** that is used to perform CPR without direct physical contact between the CPR first responder and the victim. In addition, the CPR kit will have **latex gloves** and **antiseptic wipes** to clean off the victim and the face shield. Recent guidelines for first response to cardiac events indicates that rapid chest compressions can be just as effective as chest compressions combined with the rescue breaths. However, a lack of CPR equipment should not prevent a first responder from providing life-saving CPR.

UNIVERSAL PRECAUTIONS

"Universal precautions" is a term used in the control of occupational exposure to blood-borne pathogens. The premise is that any and all blood and bodily fluids are potentially contaminated with blood-borne pathogens such as HIV or hepatitis and should, therefore, be treated as such. The universal precautions to be employed are to wear proper protective equipment when handling these items (latex gloves and eye protection) and to ensure that all surfaces contaminated with blood are properly cleaned with a bleach solution to kill any viruses that might be present. Solid waste contaminated with blood must be contained in a plastic bag to prevent any exposure to those who may be handling it.

BLOOD-BORNE PATHOGENS

Blood-borne pathogens are disease-causing agents that can be transmitted from one person to another by contact with the blood or bodily fluids of a person who is infected with them. Examples of blood-borne pathogens include hepatitis, human immunodeficiency virus (HIV), and Ebola virus. Employees may be at risk of infection by blood-borne pathogens while administering first aid or during an injury situation. They are a concern because the diseases that are transmitted by contact with blood or bodily fluid are serious with severe and sometimes fatal consequences. In addition, they do not have successful cures available, so if one becomes infected, it will impact the person for the rest of his or her life.

DISPOSING OF FIRST AID MATERIALS CONTAMINATED WITH BLOOD

During the course of administering first aid, an employer will likely generate waste materials contaminated with blood. The Bloodborne Pathogen Standard defines "regulated waste" as that which is contaminated with blood that is liquid or free-flowing. This regulated waste must be closed and contained to prevent potential contamination and exposure and should be disposed of through a licensed medical waste company. However, if there are small amounts of blood on first aid materials such as gauze pads, Band-Aids, or wound dressings that do not contain free-flowing blood, these items can be placed in a plastic bag and then placed in the regular trash and disposed of with the regular solid waste.

Safety Program Development and Implementation

Applicable Health and Safety Standards and Best Practices

Safety Training for New Hires

All newly hired employees should go through **safety orientation** on their first day. At a minimum, this training should include an introduction to workplace hazards. It should also review emergency egress routes, the location of the emergency assembly area, and emergency action procedures. There should be a review of the required **personal protective equipment** for the job, and employees should learn how to obtain new PPE supplies. If an employee will be required to wear a respirator, a **fit test** should be conducted. If complete safety training cannot be conducted on the first day of employment, the employee must not conduct unsupervised work until training is completed.

Management Practices for Handling Chemicals

Best management practices for handling chemicals include always wearing the proper **PPE** (safety glasses or goggles, chemical-resistant gloves, respirator if needed). Read the **Safety Data Sheet** and any manufacturer's literature prior to using a chemical to understand its proper use and the risks it poses. If chemicals are placed into another container, be sure to label the container. Always purchase the smallest amount of chemical needed for a job to minimize waste. Use chemicals in well-ventilated areas. Clean up spills immediately and properly dispose of the waste. Never eat, drink, or smoke when using chemicals. Always wash hands and face after using chemicals.

Waste Minimization Techniques for Construction

Waste minimization is the practice of creating as little waste as possible to minimize disposal. **Waste minimization practices** for a construction setting are based upon proper inventory management and estimation of quantities of materials necessary for a job. For example, one can calculate the amount of paint needed to perform a job using either past experience or an online calculator. This will help to minimize waste. Another waste minimization technique is substitution of less toxic materials that can be discarded as solid waste instead of hazardous waste. For example, use water-based cleaners rather than solvent-based cleaners that would have to be managed as hazardous waste.

Management Practices for Fall Protection Programs

A fall protection program should include a written plan that outlines the steps that will be taken to prevent falls. The plan should include an assessment of the types of work that will be conducted at heights, the type of fall protection equipment that will be used, and the type of training that will educate employees in equipment use. The employees must learn how to inspect the fall protection harness and how to tie the lanyard to a fixed object that will provide the fall arrest. In a manufacturing environment with routine tasks that require fall protection, tie-off points should be pre-identified and clearly marked so that employees are consistently using the correct points.

Management Practices for Confined Space Safety

A confined space is defined as one that is not designed for continuous occupancy but that a body part or entire body can enter, and that must be entered from time to time for maintenance or repair activities. The first step in **confined space safety** is to identify whether confined spaces exist, and if so, whether they are permit-required or not. A **permit-required confined space** is one that may have a hazardous atmosphere, either from low oxygen concentration or a flammable or toxic vapor present. Once the confined spaces are identified, each one must have a **sign** posted at the entrance indicating that it is a confined space, and stating whether a permit is required. Anytime someone enters a confined space, a **sentry** must be posted at the entrance to call for assistance if an employee is injured in the confined space.

COMMON COMPONENTS OF SITE-SPECIFIC SAFETY PLANS

For work that is performed at different job sites, it is important to create a **risk-based site safety plan** for each job site. The risk assessment should consider the overall process that will be conducted, as well as the site-specific risks. For example, is the area rural or urban? How close are emergency personnel in case of a fire or injury? Is emergency equipment available? What are the risks posed by use of heavy machinery? Does the site work pose risks of chemical exposure? How does the terrain or topography influence the site risks? Do any weather-related risks need to be accounted for?

Each site-specific site safety plan should list the nearest **medical facilities**, including urgent care clinics for minor injuries and hospitals for major injuries. There should be an assessment of the **personal protective equipment** needed at the job site. There should be an assessment of what other work will be conducted at the same time, and whether this will pose additional hazards. The location of power, potable water, and washing facilities should be investigated and accounted for. It should be determined if there are any extraordinary communication challenges posed by the site (for example, is it out of mobile phone signal range?).

WORKSITE ASSESSMENT OR AUDIT PROCESSES

When conducting an audit of a construction work site for compliance with health and safety regulations, examine records and work site physical conditions. **Records** include training records to show that employees have been trained in hazard communication, respiratory protection (if applicable), lockout/tagout procedures (if applicable), emergency evacuation procedures, PPE requirements, proper tool safety, and ladder safety. **Physical conditions** include evidence that safe electrical practices are followed, proper PPE is worn, and emergency equipment such as fire extinguishers is available. Routine **safety tailgate meetings** should be conducted and records of these should be maintained.

ELECTRICAL SAFETY INSPECTIONS

A routine general electrical safety inspection should be conducted at a construction site. A more in-depth inspection should be held for activities conducted by electricians. The **general electrical safety inspection** should examine all power tools to ensure the cords are intact and no wires are exposed. Power tools should be properly grounded and double insulated. Extension cords should not be used in daisy chains. There should be no exposure to live electricity. A lockout/tagout program should be in place and employees trained according to their role as *authorized* or *affected* employees. Any overhead work that could be impacted by overhead power lines must be planned and carried out to avoid touching the wires. Sources of water near electricity should be eliminated.

ROLES, RESPONSIBILITIES, AND LINES OF AUTHORITY AS THEY RELATE TO SAFETY MANAGEMENT

RESPONSIBILITY OF SUPERVISORS WITH REGARD TO EMPLOYEE SAFETY

The role of the supervisor is very important in employee safety. The supervisor is directly responsible for **communicating work instructions** and for providing employees with **on-the-job**

training in how to safely conduct the work. The supervisor enforces **policies** on safe work practices and wearing the required personal protective equipment. The supervisor is responsible for providing **disciplinary action** if the employee refuses to follow safe work practice requirements. The supervisor is also important as a **communicator with management** regarding implementation of safe procedures and recommendations for continuous improvement.

RESPONSIBILITY OF EMPLOYEES WITH REGARD TO EMPLOYEE SAFETY

An employee is responsible for his own and his coworkers' safety. An employee has the responsibility to follow all established **safe work procedures** and to wear all required **personal protective equipment**. For example, the employee must wear safety glasses if required, even if it is not preferred. The employee is responsible for communicating to the supervisor and to management if an **unsafe condition** is present and should participate in developing a solution. The employee must not take shortcuts that will endanger his and others' safety. The employee must also report to work well rested and sober so that he can give proper attention to the job and perform it safely.

RESPONSIBILITIES OF THE EMPLOYER WITH REGARD TO EMPLOYEE SAFETY

The employer has the ultimate responsibility for creating a safe work environment. The employer must design the company's **work procedures and operations** with safety in mind. Proper **job hazard analyses** should be conducted for each operational process, and hazards should be considered so they can be minimized. The preferred method is to engineer out hazards. The employer is responsible for documenting the **safe work procedures** and for communicating this information to the employees. The employer must properly **train** the employees and provide them with the proper **personal protective equipment** if needed. The employer must listen to and consider any **employee complaints** received about safe work practices, and must not retaliate against any complainants.

RECOMMENDED EQUIPMENT INSPECTION RECORDS OR LOGS

JOB SITE SAFETY INSPECTION SHEETS

A general job site safety inspection sheet should include an **inspection** for general safety best practices. For example, are employees wearing proper PPE for the jobs assigned? Are the tools in good repair? Are there slip/trip/fall hazards present in the workplace? Are any ergonomic risk factors present that can be engineered out of the work site? Are emergency egress routes established, and do employees know what to do in case of an emergency? Are fire extinguishers and emergency spill kits available? Are hazardous and/or flammable materials stored properly? If ladders are in use, are they inspected regularly?

POWER TOOL INSPECTION SHEETS

Power tools should be inspected prior to each use to ensure they are safe to operate. They should be examined to ensure they are free of **grease and oil**. One should ensure their **power source** is intact (for example, battery not cracked and power cord not frayed or wires exposed). The **guards** or fail-safe mechanisms must be present and operational. If the power tool has a **blade or drill bit**, it should be examined to ensure it is securely affixed to the tool and is not cracked or damaged. The tool should be inspected to be sure that it is **grounded and double-insulated**. If it has an electrical cord, the **third prong** should be intact. Also, ensure that the proper **PPE** is available for use with the tool (for example, if use of the tool will create flying particles, a face shield is recommended in addition to safety glasses).

INSPECTION OF FALL PROTECTION EQUIPMENT

Fall protection equipment consists of a **harness** designed to arrest a fall before it can cause serious or fatal injury. Regular inspections are important to ensure the equipment will work when needed. The **D-rings and buckles** should be inspected to ensure they are present, are not damaged with nicks or abrasions that might weaken the metal, and are not corroded. Any **stitching** on the harness must be intact and not frayed or unraveled. The **webbing material** the harness is constructed of should similarly be inspected for fraying. **Lanyards** used in conjunction with harnesses must be inspected for the same elements. Be aware that if a fall protection assembly has been used to arrest a fall it must be **retired and replaced**.

BASIC RISK MANAGEMENT CONCEPTS

BUILDER'S RISK INSURANCE

Builder's risk insurance is a policy that covers the structures under construction while they are being built. While each policy differs in its exact coverage, it is designed to protect the builder from risk of loss to the structure due to an **external event**. For example, builder's risk insurance may cover against a loss to the structure due to vandalism or an extreme natural event such as a tornado or fire. Or it may protect the contractor against a loss due to weather (for example, if the supplies already purchased by the contractor are damaged in a weather event before they can be used). This type of insurance is distinct from general business liability insurance.

GENERAL LIABILITY INSURANCE

General liability insurance for the construction trade covers losses attributed to the **contractor** in the areas of property damage, faulty workmanship, and bodily injury. For example, if a worker is driving a forklift at a job site and accidentally damages a parked car that belongs to the facility owner, a general liability insurance policy can offer coverage for the damages. Or, if a visitor is touring the job site and trips over a tool that a worker left on the floor, general liability insurance will cover that. If a client alleges faulty workmanship, a claim can be lodged against your general liability insurance policy. In addition, contractors who subcontract work are required to maintain general liability coverage to insure against losses caused by these subcontractors.

DATA GATHERING TECHNIQUES AND PROCEDURES USED IN INCIDENT INVESTIGATIONS

The primary purpose for investigating accidents is to prevent future accidents from happening. **Investigations** can also identify causes of accidents and injuries, provide evidence for legal claims and lawsuits, and help assess the amount of loss and damage. After an accident, an accident investigation should begin as soon as all emergency steps have been taken to care for the injured parties and to bring the emergency situation under control. Beginning the accident investigation quickly offers several benefits:

- Immediate investigations produce **more accurate results** because witnesses' memories are fresh and untainted.
- Immediate investigations allow the investigator to study the **accident scene** itself before it is changed.
- Immediate investigations send a message that the company **cares** about employees' safety.
- Immediate investigation demonstrates the company's commitment to discovering the cause of the accident and thus **preventing future accidents**.

Accident investigations should begin as soon as possible. The first tool an accident investigator needs is *rope* or *security tape*. Stretching rope or tape around the accident scene will help keep people out of the area, keeping the scene secure, undamaged, and unchanged. Even with rope or tape, though, evidence at accident scenes can dissipate, so the investigator should take *photos* or a *video* of the site as soon as possible. Tape measures can be used to mark where items are located and ID tags can be used for marking evidence. Investigators may also need tape recorders to interview *witnesses* about where they were and what they saw. Particular types of equipment may be needed depending on the type of accident. For example, Geiger counters are needed for radiation releases while colorimeters, sampling equipment, and clean specimen jars are needed for chemical spills.

Accident investigation can be expensive, so it is not always possible to investigate every accident. When determining whether to investigate an accident, managers need to consider the following:

- The **cost and severity** of the accident. Accidents with high losses, whether in life, injury, or property damage, need to be investigated.
- The **frequency** of the accident. If similar accidents occur frequently, they need to be investigated.
- **Public interest** in the accident. If the accident affects the community or is otherwise of special interest to the public, it needs to be investigated to provide factual information and protect the company image.
- The **potential losses** caused by the accident. If the accident may have large losses in life or property damage, it should be investigated.

TECHNIQUES FOR DETERMINING THE ROOT CAUSE OF ACCIDENTS OR INCIDENTS

Root cause analysis methodology is used to determine the most fundamental reason for a system failure or mistake that has led to an injury or equipment failure. The goal of a **root cause analysis** is to identify the central cause; this will allow the central cause to be fixed, thereby preventing a recurrence of that particular event. An effective methodology to conduct a root cause analysis is the **Five Why method**. This method uses around five *why* questions to determine the root of the problem. An example of a Five Why analysis is as follows:

- Q: Why did the valve leak oil? A: The valve had not been replaced.
- Q: Why wasn't the valve replaced? A: The workers didn't know it needed to be replaced.
- Q: Why didn't the workers know it needed to be replaced? A: No one told them.
- Q: Why didn't anyone tell them? A: There is no written replacement schedule available to them.
- Q: Why isn't there a written replacement schedule? A: No one has written one.

This process results in an identifiable task that can be undertaken to prevent a recurrence of this problem.

POST-INCIDENT/ACCIDENT REPORTING AND FOLLOW-UP PROCEDURES

INCIDENT REPORTING SYSTEM

An effective **incident reporting system** must have participation of the right people in the organization, which are those people with firsthand knowledge of the incidents being reported. In the case of injuries and near misses, diligent participation of the direct supervisors is most important. They must be properly informed of the type of incidents that need to be reported, exactly what information to report, and how to report it. In addition, the reporters must be given guidance on how the severity of the incident impacts the reporting of the incident. For example, minor incidents such as issuing a bandage for a small cut may need to be only recorded on a first aid log, whereas an injury requiring medical attention must be recorded but also requires notification of the manager on duty so that the insurance reporting and incident investigation process can begin.

A robust incident reporting system that *records* each incident without fail, and that *reports* the required information about each incident, can provide valuable information to improve future incident response. Over time, the incident log can be examined for patterns that lead to improved prevention of incidents. For example, temporal investigation may reveal that incidents occur disproportionately on the swing shift or just after break time. Incidents may be more likely to occur in one particular area; discovering these patterns can allow the organization to focus resources on prevention and response in particular areas. Analysis of incident data on leaking equipment, for example, may reveal that a certain piece of machinery develops leaks more frequently than others in the work center; the appropriate response that can be developed from this information is to put that piece of machinery on a more frequent preventative maintenance schedule. The point of tracking incidents and collecting data is to use that data as a continuous improvement tool that provides feedback on performance and guides active response.

EYEWITNESS INTERVIEWS

The purpose of interviewing eyewitnesses to accidents is to obtain information about what happened from real-time observers. Each observer witnesses and remembers different aspects of an incident. This information can be used to assist in **root cause analysis** and planning of **corrective action**. It is important to ask open-ended questions that allow the eyewitness to tell the story of what happened rather than to ask leading questions that make it seem as if the answer is already known. For example, ask "What happened after that?" to get a sequential idea of what has happened. Ask follow-up questions if a detail is unclear. In addition, the eyewitness should be asked to sign the statement provided as evidence of its veracity.

DOCUMENTATION REQUIREMENTS OF OCCUPATIONAL INJURIES AND ILLNESSES

RECORDABLE INJURIES

A recordable injury is one that must be recorded on an employer's **OSHA 300 log**. A recordable injury is a **work-related** injury or illness that results in a fatality, loss of consciousness, days away from work, or days of restricted duty work, or requires medical treatment beyond first aid. Recordable illnesses also include work-related cancer, hearing loss, cracked bones or teeth, or puncture injuries. The number of recordable injuries, benchmarked according to industry and total employee hours worked, is used to set **worker's compensation insurance rates** and is an important **safety metric** that needs to be tracked.

"FIRST AID" INJURIES

OSHA has a strict definition of what types of injuries can be considered "first aid." It is in the best interest of an employer to classify as many injuries as possible as first aid, because this reduces insurance costs. OSHA considers first aid to be a medical treatment that is **one-time** (with one follow up visit allowed) and is generally a **minor** scratch, cut, burn, splinter, or debris in the eye. Injuries that result in the doctor prescribing medication, or in days away from work or days of restricted duty, cannot be considered first aid cases.

OSHA 300 LOGS

The OSHA 300 log is a required OSHA recordkeeping element. It serves as the legal record of **recordable injuries** at a given work site. The information to be recorded includes the injured worker's name, the date of the injury, the job title, the work area the employee was in when injured, and a brief description of the injury (for example, employee experienced a lower back strain when moving heavy boxes). The log also requires the injury to be **categorized** as a case that involves death, days away from work, job transfer or restricted duty, or as an "other recordable case." If the worker experiences days away from work and/or restricted duty, the **number of days** must be tracked and recorded. If the recordable case is not an injury but an illness, one must also record the **type of illness** (skin disorder, respiratory condition, poisoning, hearing loss, or other). Be aware that **temporary agency workers** must be recorded on the OSHA log of their current location/employer when the injury occurred.

The OSHA 300A log is the **summary sheet** that accompanies the log of recordable injuries at a work site (the OSHA 300 log). On the 300A summary sheet, the **total number of recordable cases** is presented, along with the number of cases that resulted in **days away from work**, and the total number of cases that resulted in **restricted work duty**. In addition, the total number of **occupational illnesses** is recorded. The company **Standard Industrial Code** is noted, along with the total number of employee hours worked in the year. Note that this figure should not include vacation or holiday time and that it should include all workers at the site, including office workers. The OSHA 300A is an official summary record for the year of occupational injuries and illnesses and must be signed by a company official. It must be posted in an area accessible to the workforce from **February 1 through April 1** of each year.

<u>ELECTRONIC SUBMITTAL OF OSHA 300 RECORDS</u>

OSHA recently published a rule requiring electronic submittal of OSHA 300 records to OSHA by certain industries and businesses. The rule may be found in 29 CFR §1904.41, **Recording and Reporting Occupational Injuries and Illness**. Businesses with more than 250 employees at any time during the previous calendar year that are required to keep OSHA 300 and 300A records must electronically submit the **Form 300A information** by March 2 of the next year (starting in 2019; in 2017 and 2018 the reporting deadlines were different). Businesses in specified industries with 20 – 250 employees must also report. The specified industries are diverse industries such as manufacturing, agriculture, utilities, construction, various types of stores, trucking, warehousing, nursing care, etc. These industries are listed in Appendix A to 29 CFR §1904 Subpart E.

OCCUPATIONAL ILLNESSES

An occupational illness (as opposed to an injury) is an illness or condition that can be attributed to an **occupational exposure in a workplace** through a physical, chemical, or biological cause. Examples of occupational illnesses are poisoning caused by exposure to a toxic chemical (such as exposure to lead dust or fumes), or hearing loss caused by overexposure to noise. Other examples that could occur in a construction setting are occupational lung diseases such as asbestosis (from exposure to asbestos) or silicosis (from exposure to silica dust). Contact dermatitis (skin irritation or inflammation) can be caused by exposure to a chemical or to latex in gloves. Carpal tunnel syndrome or other repetitive use disorders are also occupational illnesses.

DART

The DART statistic is the number of injury and illness cases with "**D**ays **A**way and/or **R**estricted Duty **T**ime" in a given year. The statistic is normalized by the number of hours worked and refers to the number of **recordable incidents** per 100 full-time workers. The calculation is as follows:

DART Rate = (Total Number of DART Incidents) x (200,000) ÷ (Total Hours Worked)

The 200,000 figure normalizes the statistic to 100 full-time workers and allows the DART rate to be compared across industries and workplaces of different numbers of employees. A general rule is that the **DART rate** should be less than the **Total Case Incident Rate**, which indicates that only a portion of recordable cases was severe enough to result in restricted duty or lost work time.

TCR

The TCIR is the **Total Case Incident Rate** (sometime called the IR, or Incident Rate), and represents the total number of recordable injuries and illnesses per 100 employees. The calculation is as follows:

TCIR Rate = (Total Number of Recordable Cases) x (200,000) ÷ (Total Hours Worked)

The TCIR is tracked by the **Bureau of Labor Statistics** for a wide variety of industries and published by **Standard Industrial Classification code**. This is a useful tool in benchmarking a given company's safety performance against others in the same industry.

DOCUMENTING OCCUPATIONAL INJURIES

After ensuring that an injured worker has the proper medical treatment, it is important to follow a protocol to properly **document** the injury, both for legal and insurance purposes. A **statement** should be taken from both the injured worker and any eyewitnesses regarding what happened. A **root cause analysis** should be done to determine how processes or procedures can be altered to prevent recurrences. Records of all **medical visits** should be maintained. If a worker is on **restricted duty**, a formal record should be kept, stating whether restricted duty is offered to the worker, and the conditions of acceptance. If the injury requires reassessment of a **job hazard analysis**, this record should also be maintained.

REPORTING SEVERE INJURIES

A severe injury is defined as any occupational injury that results in an **amputation, loss of an eye, inpatient hospitalization**, or a **fatality**. Work-related fatalities must be reported to OSHA within **eight hours** and other types of severe injuries must be reported within **24 hours**. An amputation is the loss of any limb or appendage, even the tip of the finger. The company must report the name of the affected employee, the location the injury occurred, the company name, the type of event, a brief description of the incident, and the contact person who can be reached for further information. OSHA concentrates investigation efforts on workplaces with severe or serious injuries, and generally follows up with a **site visit** to verify information and compliance status related to the severe injury event.

DETERMINING WHETHER AN INJURY IS WORK-RELATED

The regulation that states the definition of "work-related" is found in 29 CFR §1904.5(a), which states that an injury or illness is work-related if "an event or exposure in the work environment either caused or contributed to the resulting condition." Sometimes this is obvious; for example, if an employee is using a saw and is cut by the saw, this is obviously work-related. In the case of musculo-skeletal strains, it is not always so clear. However, if a worker claims that an injury is work-related, it is **assumed to be so** unless proven otherwise. In other words, the worker is given the benefit of the doubt. Even if the worker's off-hours activities may have contributed to the injury (for example, participation in a sports league), an injury will be designated work-related if it could have been caused by the work.

MAINTAINING OSHA LOGS FOR MULTIPLE JOB SITES

OSHA regulations regarding recordkeeping are found in 29 CFR Part 1904. The rules for recording injuries that may occur at **remote job sites** are relevant in the construction industry; for example, an electrical contractor may have employees performing work at several job sites simultaneously. For employees that work at a given job site **less than one year**, the employer may maintain one OSHA 300 log and record all recordable injuries on the one log. If an employee is working at a given location for **longer than one year**, individual logs must be maintained for each site. The log may be maintained at a central location or corporate headquarters as long as it can be provided to regulators upon request in a reasonable length of time.

Leadership, Communication, and Training

Available Training Delivery Methods and Instructional Materials

Classroom Training

Classroom training is often used for safety training, but it has its pros and cons. On the **positive side**, it offers a chance for workers to dedicate their attention to the task without distractions. It allows for questions to be asked to clarify concepts. Some types of information are best presented in a classroom; for example, hazard communication and toxicology of chemical exposures are best presented in a classroom setting. On the **negative side**, it can be difficult to make the training class lively and engaging and the participants may get sleepy. It may not be interactive enough. Some people do their best learning by doing and not by listening to another person give a presentation.

On-the-Job Training

On the job training refers to training given outside of the classroom setting, in the work area. It is usually for teaching the actual mechanics of a production job; however, it is also a great opportunity to present **risk and hazard information** in a real-world setting. A positive reason to do on the job training is to provide an opportunity for **interaction and hands-on training**. This can help solidify the concepts and make the training more relevant. A negative aspect of doing on-the-job training is that it is not well-suited to providing information about **academic concepts** such as chemical exposures or hearing conservation. It is also not a good method for large groups or noisy work areas.

Online Training

Online, web-based training classes are useful because they are available any time and do not require scheduling the physical space and the trainer. They can be very **cost-effective**, and can be tailored to the specific topic to be delivered. Programs purchased from a training program vendor can also keep track of the **training records**. They can also have built-in **competency assessments**. On the negative side, they provide little **interaction** and may be difficult for some people to pay attention to. They do not provide an opportunity for the student to ask questions and get **clarification** on concepts that aren't understood. Some people may not be familiar with how to use a computer, depending on literacy levels and background.

Training Delivery Mediums and Technologies

Safety training can be difficult and tedious, but it is critical for worker health. Classroom training is beneficial because content can be tailored to the audience and work environment. It also allows for personal interaction and hands-on training activities that enhance learning. The disadvantages include the difficulty of scheduling a class that can accommodate everybody who needs to take the class. Usually multiple times and locations must be offered. Online training classes are an attractive option because they can be completed whenever the individual has the time. However, they tend to be generic and not tailored to the audience or specific work environment. Sitting in front of a computer terminal is often boring and it is more difficult for employees to understand and retain information delivered this way.

Choosing Appropriate Training

A variety of training techniques should be used to teach and reinforce concepts. The best training programs use a combination of all three methods. Traditional classroom lecture-style training is useful when the topic is academic in nature. For example, it can be helpful to explain lockout/tagout regulations and procedures in a classroom setting. Demonstration training is best when the topic involves physical manipulation of an object. An example of demonstration training is teaching the use of a respirator, how to disassemble and maintain it. On-the-job training refers to training in specific tasks and occurs in the actual work environment. Written reference material and job aids should be available to reinforce the proper performance of each task. In the example of lockout/tagout above, on-the-job training should be conducted with each piece of equipment to show employees where the lockout points are, and to demonstrate proper isolation of the energy source.

LEARNING STYLES

Although there are many theories about learning styles in the educational field, one popular theory that can be applied to safety training classifies learning styles into four types (acronym: VARK):

- **Visual** – This type of learner retains information best by seeing information presented visually, such as reading it for themselves as opposed to listening to a presentation. They can also benefit from information presented in charts or graphs.
- **Auditory** – This type of learner responds to hearing information, such as a lecture or oral presentation.
- **Reading/Writing** – This type of learner responds to reading, followed by note taking or re-writing information. The act of taking notes reinforces the concepts and helps the learner to retain knowledge.
- **Kinesthetic** – A kinesthetic learner learns best by doing an activity related to the concepts. For example, this type of learner would respond to a hands-on exercise of how to disassemble and reassemble a respirator as a way of learning how it works, rather than viewing a PowerPoint presentation on the topic.

The best types of safety training classes integrate aspects that will appeal to all of these learning styles, and reinforce concepts using multiple methods.

APPROPRIATE HUMAN BEHAVIOR MOTIVATION METHODS AND TECHNIQUES

BEHAVIORAL-BASED SAFETY

Behavioral-based safety is a system of improving a safety culture and safety performance that concentrates on recognizing and changing employees' unsafe behaviors. Although it is preferable to improve safety by engineering out hazards, it remains a fact that many on-the-job injuries result from individual employees doing unsafe acts, despite policies and procedures that are put in place to ensure a safe workplace. For example, an employee may lift an object without employing safe lifting procedures or may elect to use a tool without donning the appropriate personal protective equipment (PPE). Implementing a behavior-based safety program aims to identify unsafe acts and uses peer observations and corrective actions to improve safety performance and adherence to safe workplace policies.

KEY ELEMENTS OF SUCCESSFUL IMPLEMENTATION: Behavioral-based safety must first start with a sound assessment of the types of behaviors that are important in a given work environment to reduce the incidence and severity of injuries. Past data on injuries, first aid, and near misses should be evaluated to determine the common unsafe acts that could lead to incidences. These are the behaviors that can be concentrated upon first. Second, teams of safety observers must be recruited for participation in the behavior-based safety program. The line-level workers must be trained in how to make observations of behavior for unsafe acts, and the workers who have performed the unsafe acts need to be notified and corrected on the spot but not in a punitive way. There should also be ongoing feedback to the entire team on the aggregate behavioral observations and how the team is improving over time so that the benefits of the system can be observed.

POTENTIAL PITFALLS: There are potential pitfalls in the execution of a behavioral-based safety program. The successful implementation requires involvement and participation from line-level workers to make the safe behavior observations and to provide feedback to their coworkers. If proper training and support are not provided, the observations made will not be useful. In addition, the team has to understand the potential benefits of improving safety behaviors and cannot be made to feel that they are being blamed for injuries that are out of their control. There should be a definite and observable connection between the safe behaviors observed and the types of injuries and incidents being reduced through improved behavioral performance.

COMMUNICATION STRATEGIES

There are several ways to communicate safety information to workers. One is **written information** in the form of work instructions or safety guidelines. Another is to post reminder signs with **pictures** to communicate the information. A **company newsletter** can be used to publish reminders or other information. **Safety tailgate meetings** can provide a valuable time for two-way communication. **Posters** can be used to provide safety reminders. **Video monitors** can show a PowerPoint presentation on a continuous loop; this is effective to use in a break room or in a situation in which employees are standing in line and are therefore a captive audience.

Clear communication is critical for safety. Job procedures, PPE requirements and emergency procedures should be communicated in training courses, reviewed in safety meetings, and refreshed and reinforced regularly. To help reinforce these training methods, communication aids should be created and posted in the work areas to serve as a reference and a reminder.

Job procedures – should be written in a way that communicates in words the steps needed to complete a given job safely. The procedure should also use pictures and illustrations as much as possible so that communication is visual as well as written. These can be posted at the point of use as a reference and reminder.

Personal Protective Equipment – required PPE should be listed and posted near the equipment along with a photograph or illustration for clarity.

Emergency Procedures – Emergency exit routes should be clearly indicated on facility evacuation maps that are posted in work areas. It is also helpful to have evacuation routes drawn on the floor in reflective paint. Emergency exits must be indicated with lighted signs. An emergency procedures checklist should be readily available in the work area that lists any equipment that should be de-energized during emergencies. Also, someone must be designated to ensure everyone has evacuated a given area and accounted for in the assembly area.

WHEN TO CONSULT WITH EQUIPMENT MANUFACTURERS, SUPPLIERS, OR SUBJECT MATTER EXPERTS

Manufacturers and suppliers can be valuable sources of safety information. They know best how their products work and which product is best for a given hazard. For example, glove manufacturers should be consulted when selecting gloves for particular hazards. They can provide information on chemical and physical resistance of the gloves and suggest alternative products that may be more cost-effective for your application. When you are purchasing a new piece of equipment, consulting the manufacturer in advance to determine potential safety risks of the process is essential. The manufacturer will often be able to provide safety and training materials. Forklift manufacturers provide training materials for their products and usually have personnel that can come on-site to conduct the training. Consulting with manufacturers of safety equipment such as machine guards can provide valuable information on the right solution for your machine guard hazard. Although consulting manufacturers does not always yield a solution, it is an avenue that should be explored.

There are multiple subject matter experts who can be consulted to promote optimum safe practices. For example, one should always consult the maintenance and engineering staff for input on mechanical and machine hazards. They can provide information on the operation of various machines and the potential hazards, what hazardous chemicals may be used in maintenance activities, and use of emergency shut-off mechanisms. Other subject matter experts to consult are industrial hygienists for advice on exposure limits to determine appropriate levels of respiratory protection. Worker's compensation insurance providers may be able to provide invaluable information to help manage worker's compensation claims to minimize costs while providing proper care for employees injured on the job. In order to minimize ergonomic hazards and prevent injuries from ergonomic stresses it is important to consult an ergonomics expert.

INFORMATION CONFIDENTIALITY REQUIREMENTS

EMPLOYEE HEALTH INFORMATION

Employers are not entitled to all the health and safety information of employees. There are legal privacy protections under the Health Insurance Portability and Accountability Act (HIPAA) that protects employee's health information if it is not connected to a work-related injury, request for sick leave, or provision of health insurance. Health information provided to the employer must be authorized by the employee. Employers are authorized to receive work readiness health information for employees that have work-related injuries. These are not actual medical records but are the physician's assessment of an employee's work restrictions or prognosis. Alcohol and substance abuse records are not considered occupational health records. There are other health-related records employers are obligated to retain. For example, records of employee exposures to toxic materials must be retained for thirty years. This includes industrial hygiene monitoring records and the results of any biological monitoring such as blood lead levels or cadmium in urine levels. Results of audiometry must be similarly retained. The purpose of this requirement is that certain adverse exposure outcomes may take years to manifest (such as hearing loss or cancer) and exposure records need to be available in case there is a possibility that the problem is work-related.

TRADE SECRETS

A trade secret is business information that the business does not want disclosed to customers or competitors but is not protected by any formal procedure such as a copyright. To be considered a trade secret, the information must not be generally known, it must have commercial value because it is secret, and the organization must have taken steps to keep the information confidential. Steps that organizations must take to keep trade secrets include identifying what the information is and to whom it will be disclosed. Steps must be taken to ensure written documents are stamped or marked confidential and that there is an assured destruction method available to discard written materials (such as shredding). Computer files that contain confidential information must be password protected, and the electronic file storage systems must be designated and access granted only to those who need to know. There should be a documented procedure for granting computer file access. Employees must be given training on the type and nature of trade secret information and the importance of securing this information.

BCSP Code of Ethics

The Board of Certified Safety Professionals (BCSP) has established a code of ethics and professional conduct that must be followed by individuals who are awarded certificates by this organization. The first two standards of this code promote the need for certificate holders to support and promote integrity, esteem, and influence of the safety occupation. The code prioritizes human safety and health as the top concern in any scenario. Additional focus should also be given to environmental safety and the protection of property. Each of these priorities can be promoted by safety professionals in warning people of hazards and risks. These standards also include the promotion of honest and fair behavior toward all individuals and organizations and the avoidance of any behavior that would dishonor the esteem or reputation of the safety profession.

Standard 1 focuses on the primary responsibility of a Construction Health and Safety Technician (CHST) to protect the safety and health of humans. A specific component of this responsibility is the need to notify appropriate personnel, management, and agencies regarding hazardous or potentially hazardous situations. An example might be a scenario in which a CHST is analyzing company records of employee exposures to potentially hazardous materials on a consultant basis. The CHST discovers a calculation mistake in transferring readings from a personal monitor into a formula that calculates the daily exposure rate. This mistake has resulted in underestimating the exposure rate by as much as 50%. The CHST immediately notes these findings in an urgent communication to the company. He then follows up with the company on correcting the calculation errors immediately for present and future employees and determining actions taken for employees that might already have received hazardous levels of exposure.

Standard 2 specifically deals with the CHST's character. The CHST should "be honest, fair, and impartial; act with responsibility and integrity…" The CHST has to balance the interests of all involved parties, and represent the profession in all business dealings.

Standard 3 focuses on appropriate public statements and contact. This standard notes that honesty and objectivity are required in all communications and statements. Construction Health and Safety Technicians (CHST) should only make statements related to areas or situations about which they have direct expertise.

Standard 4 of this code focuses on the professional status and actions of CHST's. A CHST should only engage in projects or activities for which they are highly qualified in terms of knowledge, training, and experience. Additionally, ongoing education and professional advancement activities should be pursued by all CHST's to maintain and improve knowledge and skill level as well as to maintain certification. The Board of Certified Safety Professionals supports annual conferences that enable CHST's to receive valuable training and education while accumulating required points to keep their certification current.

Standard 5 focuses on integrity and honesty in the presentation of professional qualifications. This includes areas such as education, degrees, certification, experience, and achievements. Not only must CHST's be careful to clearly and honestly state all qualifications, but any exaggeration or misrepresentation by omission must also be avoided. When applying for jobs, providing references, or testifying in court, a CHST must not lie about their employment history, professional relationships, or professional qualifications and experience. An CHST with knowledge about violations of this standard should report the information to the Board of Certified Safety Professionals.

Standard 6 specifies that the CHST should avoid all conflicts of interest in order to maintain the integrity of their profession.

Standard 7 focuses on avoiding discrimination and bias. It is specifically noted that CHST's must not discriminate or demonstrate bias based on gender, age, race, ethnicity, country of origin, sexual orientation, or disability. These areas are also regulated by local, state, and federal agencies enforcing Civil Right and other related anti-discrimination legislation.

Standard 8 of this code promotes the need for CHST's to be involved in community and civic events and use their professional qualifications to promote safety within their own community. Examples of this include instructing public safety personnel in handling hazardous materials, teaching public safety classes in areas of fire prevention, home safety, and accident prevention. CHST's can work cooperatively with organizations and agencies already working in the community such as the Red Cross, fire departments, and police departments.

A Construction Health and Safety Technician (CHST) should adhere to the impartiality principle in all interactions and engage the advice and support of supervisors in determining appropriate actions. One area of concern is to avoid conflicts of interest or promoting oneself beyond actual expertise or competence. For example, if asked to testify in a lawsuit that revolves around an incident involving a former employee of a company affiliated with the CHST's employer, the CHST should defer to avoid conflict of interest. Another example would be a CHST who is asked to testify to the appropriateness of safety procedures related to radioactive materials. Although the CHST received general instruction in this area in a course, he has no direct experience. He should decline to avoid presenting himself as an expert in that area.

The Code of Ethics and Professional Conduct for Construction Health and Safety Technicians (CHST) notes that the primary responsibility of a safety professional is to safeguard the health and safety of humans as well as providing for the safety of the environment and property. In a typical construction site, numerous organizations, individual, and entities are working jointly to complete the project. Ultimately, each organization or company bears responsibility for making sure its workers are well trained and have a safe work environment. A main function in ensuring a safe worksite is to include safety stipulations in the contracts of all involved individuals or organizations.

A CHST must earn at least 20 points every five-year period, based on continuing education and professional development in order to retain their certificate.

CHST Practice Test

1. Which of the following is NOT considered a significant factor in determining the outcome of an electrocution event?

    a. The level of electrical current passing through the body
    b. The amount of electrically resistant material present in work boots and work gloves
    c. The total amount of time an electrical exposure occurs
    d. The actual path through the body in which the electrical current flows

2. A Noise-Hazard Control Program (and associated Hearing Conservation Program) is required under OSHA regulation 29 CFR 1910.95 (Occupational Noise Exposure) where workers may be exposed to noise levels in excess of an eight-hour time-weighted average of _____.

    a. 60 decibels (dB)
    b. 75 decibels (dB)
    c. 85 decibels (dB)
    d. 95 decibels (dB)

3. As per 29 CFR 1910.1001, the allowable eight-hour exposure limit concentration to asbestos fibers is not to exceed which of the following?

    a. 0.1 fibers per cm^3 of air
    b. 0.25 fibers per cm^3 of air
    c. 10 fibers per cm^3 of air
    d. 100 fibers per cm^3 of air

4. Per OSHA regulations 29 CFR 1910.25-1910.27, the maximum length a single ladder or a single section of ladder may be is _____.

    a. No greater than 20 feet
    b. No greater than 30 feet
    c. No greater than 36 feet
    d. No greater than 48 feet

5. Which of the following is NOT stipulated as a hazardous waste material attribute to which the U.S. Environmental Protection Agency extends a designated waste code (e.g., #D001), as per 40 CFR 261, Subpart C?

    a. Toxicity
    b. Corrosivity
    c. Mutagenicity
    d. Ignitability

6. In a construction environment, which of the following types of PPE is most suitable for protecting the eyes against airborne dust?

    a. Safety glasses
    b. Safety goggles
    c. Welding goggles
    d. Industrial optic-shields

7. At what minimum frequency must electrical-insulating gloves be tested to ensure they adequately protect workers against electrical shock?
   a. Every six months
   b. As deemed appropriate by visual inspection
   c. Annually
   d. Biweekly

8. The most useful tool that CHSTs can potentially use for evaluating system safety levels at smaller-scale worksites is a(n):
   a. Failure modes and effects analysis (FMEA)
   b. Event tree
   c. Root-cause analysis
   d. Job safety analysis (JSA)

9. If 25 or more employees are to be working underground at the same time, which of the following safety procedure(s) **must** be in effect?
   a. Two separate, independent telephone lines to the underground must be active and operational.
   b. Supplied-air volumetric flow rates must be cross-checked and verified prior to entry.
   c. There must be an accountability-journeyman present underground at all times.
   d. Two separate rescue teams must be available.

10. If a five-foot deep trench is dug in clay material, what angle must the trench wall not exceed?
    a. 53 degrees
    b. 63 degrees
    c. 72 degrees
    d. 22.5 degrees

11. "Lifting index" (LI) is calculated per which of the following?
    a. Load weight × recommended weight limit
    b. Load weight ÷ recommended weight limit
    c. Load weight ÷ weight of worker
    d. (Load weight ÷ worker age)^2

12. "Flash point" is most accurately defined as which of the following?
    a. The highest temperature at which a vapor will ignite in air
    b. The lowest concentration of vapor that will ignite in air
    c. The temperature at which a vapor and liquid phase are in a state of equilibrium
    d. The lowest temperature at which a flammable liquid can form an ignitable mixture in air

13. Which of the following does a mobile-crane operator NOT need to be aware of in order to ensure a safe lift?
    a. Whether the outriggers are extended or retracted
    b. The angle of the boom
    c. The width of the jib
    d. Whether the tires are fully inflated

14. To what part of the body does a body harness typically NOT distribute any significant fall-arrest force?

   a. Shoulders
   b. Lower back
   c. Pelvis
   d. Thighs

15. "Recommended weight limit" (RWL) is most accurately defined as which of the following?

   a. The weight that healthy workers could lift for up to eight hours without causing injuries
   b. Fifty pounds for healthy male workers and 30 pounds for healthy female workers
   c. The maximum weight associated with a body mass index of 25 (threshold of "overweight")
   d. The weight at which a back-support apparatus should be worn

16. The Heinrich "incident to injury ratio" model states that for every 330 accidents, _____ result in no injuries, _____ cause minor injuries, and _____ cause(s) major injuries.

   a. 230, 99, 1
   b. 250, 70, 10
   c. 275, 50, 5
   d. 300, 29, 1

17. Which of the following is regarded as the "three E's of safety"?

   a. Enlightenment, education, execution
   b. Education, evolution, execution
   c. Engineering, education, enforcement
   d. Education, execution, excellence

18. Total case incident rate (TCIR) is calculated via which of the following?

   a. Number of recordable injuries per year ÷ total hours worked
   b. (Number of recordable injuries per year × 200,000) ÷ (total hours worked)
   c. (Number of recordable injuries per week × 40) ÷ (total hours worked)
   d. (Number of recordable deaths ÷ number of recordable injuries) × (total hours worked)

19. Which type of incident usually accounts for the highest fatality rate in the construction industry?

   a. Exposure to harmful substances
   b. Falls
   c. Fires and explosions
   d. Contact with objects and equipment

20. According to OSHA regulation 29 CFR 1926.651, a stairway, ladder, ramp, or other safe means of egress shall be located in trench excavations that are ___ feet or more in depth.

   a. Four
   b. Five
   c. Six
   d. Seven

21. Excavation cave-in protection is ALWAYS required when which of the following conditions are met?

   a. The excavation mainly comprises gravel or sand, and is greater than three feet in depth.
   b. The excavation mainly comprises clay or silt, and is greater than five feet in depth.
   c. The excavation is greater than four feet in depth, regardless of material.
   d. The excavation employs heavy equipment to remove excavated material.

22. What level of electrical current is enough to potentially result in death?

   a. 0.1 milliampere
   b. 1 milliampere
   c. 15 milliamperes
   d. 75 milliamperes

23. Which of the following types of construction equipment can NOT make inadvertent contact with overhead power lines?

   a. Backhoes
   b. Scaffolding
   c. Raised dump truck beds
   d. Bulldozers

24. Which of the following is NOT a recommended course of action for preventing falling-object injuries to workers below a scaffold?

   a. The area beneath the scaffold should be barricaded to the extent practicable.
   b. Panels or screens should be erected if loose objects are stacked higher than toe boards.
   c. A safety buffer zone of 12 feet around the perimeter of a scaffold, for every 10 feet of elevation, should be employed while the scaffold is in use.
   d. A canopy or net should be installed below the scaffold to catch or deflect falling objects.

25. Per OSHA regulation 29 CFR 1926.502, when a 200-pound load is applied in a downward direction to the top edge of a guardrail, it shall not deflect to a height any less than _____ above the walking or working level.

   a. 28 inches
   b. 34 inches
   c. 39 inches
   d. 46 inches

26. Which of the following is NOT a type of welding?

   a. Modulus welding
   b. Resistance welding
   c. Arc welding
   d. Oxygen-fuel gas welding

27. Which of the following is a common test used to check for cracks in a grinding wheel?

   a. Visual inspection
   b. Non-destructive assay
   c. The "ring test"
   d. The "lubricated disc" test

28. Which of the following soil classifications is correct?
   a. Type "A" – Cohesive soil with a low compressive strength
   b. Type "B" – Cohesive soil with a moderate compressive strength
   c. Type "C" – Cohesive soil with a high compressive strength
   d. Type "D" – Cohesive soil with a low compressive strength

29. Electrical "bonding" is best defined as:
   a. The attraction of polarized charges within an open circuit
   b. The transfer of electrons from one atom to another through an ionic bond
   c. The process of not being able to detach from an electrified object during an electrocution event
   d. Connecting two or more conductive objects with a conductor to protect against electrocution.

30. According to OSHA regulation 29 CFR 1910.242, the compressed-air pressure limit for cleaning operations shall not exceed _____ pounds per square inch.
   a. 20
   b. 25
   c. 30
   d. 40

31. Which of the following materials or objects is NOT typically used for electrical grounding?
   a. Ventilation ducts
   b. Cold water pipes
   c. Building steel
   d. Ground rods

32. Anchor points for personal fall arrests should be capable of holding _____ pounds per person.
   a. 300
   b. 1,000
   c. 5,000
   d. 7,500

33. Cranes and derricks must be clear of power lines by at least ____ feet for voltages up to ____ kV.
   a. 5, 10
   b. 5, 25
   c. 10, 25
   d. 10, 50

34. Whose ultimate responsibility is it to ensure that only well-maintained and operable hand and power tools are utilized on a job site?
   a. The employee
   b. The employer
   c. OSHA
   d. Contracted labor unions

102

35. Which of the following is NOT an OSHA-mandated standard for good housekeeping?
    a. Containers shall be provided for the collection and separation of waste, trash, and oily debris.
    b. Containers used for caustics, acids, and harmful dusts must be equipped with covers.
    c. Daily clear scrap lumber and associated debris from passageways and stairwells.
    d. Combustible scrap and associated debris shall be removed at regular intervals.

36. Under current laws for workers' compensation, which of the following is NOT categorized as a potential injury category?
    a. Partial
    b. Bilateral
    c. Total
    d. Permanent

37. When the body's temperature regulation system fails and sweating becomes inadequate, which of the following heat-related conditions is likely at hand?
    a. Heat exhaustion
    b. Heat stroke
    c. Dehydration
    d. Heat syncope

38. Which of the following is generally NOT regarded by OSHA as a "control method" for preventing work-related injuries?
    a. Adherence to the Fair Labor Standards Act (FLSA)
    b. Personal protective equipment
    c. Engineering controls
    d. Administrative controls

39. The ISO 14001 environmental management system is built upon which of the following models?
    a. Strategize, conduct, verify, conclude
    b. Scope, order, efficiency, accountability
    c. Plan, do, check, act
    d. Scope, evaluate, reassess, learn

40. Which of the following is NOT a typical characteristic of a "confined space"?
    a. It is large enough for an employee to temporarily enter and perform an assigned task without duress
    b. It has limited or restricted entry options.
    c. It has a continuous air supply.
    d. It is designed for a brief period of human occupancy.

41. Which of the following scenarios is indicative of a hazardous atmospheric environment?
    a. An atmospheric carbon dioxide concentration level of 100 ppm.
    b. An atmospheric oxygen concentration at 22 percent.
    c. An inert gas exists at a concentration level of 200 ppm.
    d. A flammable gas exists at a concentration of 25 percent of its LFL.

42. Properly measuring for excessive heat or cold in the workplace (including workers themselves) could possibly entail any or all of the following EXCEPT:

   a. Measurement of workplace temperature and wind speed
   b. Requesting that workers initially perform scouting surveys of an area prior to performing measurements
   c. Verifying proper operation of pertinent HVAC systems
   d. Monitoring worker vital signs and general functionality (e.g., pulse rates, stress levels, possible dehydration)

43. Combustible liquids that are not stored within the confines of a "designated" outside storage area must:

   a. undergo a canister filter change-out at least semiannually
   b. be double-anchored to the ground or other permanent structure
   c. be out of direct sunlight at all times
   d. be completely surrounded by a curb structure or enclosure that is equal to or greater than six inches in total height

44. Per OSHA's 29 CFR 1904.6 (Recordkeeping Standard), for how long must a document such as the OSHA 300 Log be retained?

   a. Five years following the end of the calendar year that the record covers
   b. Seven years following the end of the calendar year that the record covers
   c. Six months following the end of the calendar year that the record covers
   d. Two years following the end of the calendar year that the record covers

45. The OSHAS 18000 series of occupational health and safety management system standards have the overarching goal of:

   a. Proactively identifying and responding to risk before an accident occurs
   b. Advising workers on safe work practices
   c. Summarizing all construction site standards as outlined in 29 CFR 1910
   d. Providing strict managerial and supervisory response guidelines in the event of an accident

46. A cut-off wheel that has a diameter in the range of 6-12 inches can cut a material with a maximum thickness of:

   a. 1/8 inch
   b. 1/4 inch
   c. 1/2 inch
   d. 3/4 inch

47. Per OSHA 29 CFR 1910.1025, the action level for lead (Pb) exposure is 30 micrograms per cubic meter of air, averaged over an eight-hour period. What is the permissible exposure limit?

   a. 15 micrograms per cubic meter of air, averaged over eight hours
   b. 30 micrograms per cubic meter of air, averaged over eight hours
   c. 50 micrograms per cubic meter of air, averaged over eight hours
   d. 300 micrograms per cubic meter of air, averaged over eight hours

48. Training construction employees on the OSHA Hazard Communication Standard is required:
   a. Semi-annually, with initial training being performed within one month of the employment start date
   b. When a new hazardous material is introduced into the workplace
   c. For a worksite to potentially qualify for VPP-Star status
   d. After an incident occurs with a hazardous material

49. A "tagout" process must be employed when an energy-isolation device is not able to be locked out. Which of the following is a typical requisite when implementing the "tagout" process?
   a. Tags must be normalized to a consistent size and shape.
   b. Tags must be either red or orange in color.
   c. Tags must always be affixed in at least two places by a metallic clip.
   d. Tags must display the term "Hazard" or "Electrical Hazard" on the outward-facing side.

50. Standards for safety glasses are set by which of the following?
   a. The American Society for Testing and Materials
   b. The Occupational Safety and Health Administration
   c. The American National Standards Institute
   d. The National Institute of Standards and Technology

51. Under which of the following incident causal factors would "inadequate/substandard work procedures" apply?
   a. Substandard management or oversight practices
   b. Shortcutting
   c. Poor safety culture
   d. Substandard training

52. The construction health & safety technician can best identify potential workplace hazards via:
   a. Performing a workplace fault-tree analysis.
   b. Performing a failure-modes-and-effects analysis
   c. Performing a rigorous worksite analysis
   d. Analyzing recent accident and injury trends in the workplace

53. What type of OSHA violation can result in a fine of up to $250,000 if an employee death occurs?
   a. Willful violation
   b. Contributory negligence
   c. Repeated violation
   d. Errors and omissions

54. When performing safety calculations, especially in regard to potential falls, which of the following correctly depicts the acceleration of gravity constant?
   a. 9.8 ft/sec^2
   b. 32.2 ft/sec^2
   c. 32.2 m/sec^2
   d. 9.8 ft/sec

55. Using the Pythagorean Theorem, if a CHST wanted to find the length of "a" knowing that "b" = 12 ft and "c" = 13 ft, what would be the correct value for "a"? (NOTE: drawing is not to scale)

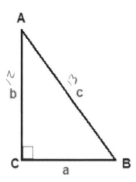

a. 1 ft
b. 2.2 ft
c. 5 ft
d. 17.7 ft

56. Which of the following is typically included in a worksite analysis that addresses potential instances of workplace violence?

   a. Manager interviews
   b. A worksite security evaluation
   c. Procedural controls
   d. Metal detector inspection

57. A safety intervention is defined as an action to modify how work is conducted with the ultimate goal being to enhance worker safety and health. Which of the following would NOT be characterized as a prototypical "safety intervention"?

   a. Safety program participation initiatives
   b. Enhanced training requirements
   c. Design modifications
   d. Lost-time trend evaluations

58. Per OSHA 29 CFR 1910.23, a "floor-hole" is dimensionally defined as:

   a. Between ½ inch and 6 inches
   b. Anything greater than 8 inches
   c. Between 1 inch and 12 inches
   d. Anything greater than 14 inches

59. Per ANSI Standards (ANSI/ASSE A1264.2 – Slip Resistance on Walking/Working Surfaces), a coefficient of friction close to "0" is regarded as:

   a. Not slippery
   b. Somewhat rough
   c. Very rough
   d. Extremely slippery

60. ASTM D 120-09 requires that rubber-insulating gloves for use in live electrical work exceeding _____ must be _____.

    a. 100 volts / Type-II
    b. 50 volts / Type-I
    c. 200 volts / Type-I
    d. 500 volts / Type-III

61. Which of the following is NOT typically considered a potential hazard for vehicles such as bulldozers, backhoes, and forklifts?

    a. Overloading
    b. Operator visibility
    c. Carbon monoxide overexposure
    d. Tipping

62. Which of the following is NOT an example of "point-of-operation" guarding?

    a. Ring guards
    b. Enclosure guards
    c. Interlocked guards
    d. Fiscal guards

63. At a construction worksite, which of the following is NOT a system or device that is typically employed for limiting falls?

    a. Safety nets
    b. Pitons and safety cables
    c. Harnesses
    d. Catch platforms

64. Which of the following products is often utilized in the workplace for reducing hazards associated with standing for long periods of time?

    a. Anti-fatigue mats
    b. Arch supports
    c. Foam underlays
    d. Work boot risers

65. Per OSHA protocol, the use of fall-protection equipment must be used when a construction worker is at least _____ above the ground.

    a. 15 feet
    b. 10 feet
    c. 8 feet
    d. 6 feet

66. Which of the following is NOT generally considered a control for storing materials safely?

    a. Use of crossties
    b. Use of retaining walls
    c. Employing a "stepping-back" layout for materials stacked in several rows
    d. Storing pipes and bars in racks that face main aisles.

67. During a rigging inspection, if the throat opening of a hook has increased in excess of _____ percent of the original throat opening, it should be replaced.
    a. 10
    b. 15
    c. 22.5
    d. 30

68. An emergency plan should always include procedures for sheltering in place, which need(s) to acknowledge which of the following?
    a. Where occupants should gather upon a shelter-in-place event
    b. How to evacuate an area or structure when a shelter in place is executed
    c. How to shut down all ventilation systems and elevators
    d. How to inform all area occupants that an emergency has occurred, and what protocol they need to follow

69. What is the definition of an "on-scene coordinator"?)
    a. A person who is in charge of coordinating various agencies and departments to ensure that all needs are responded to during and after an event.
    b. The lead federal- or state-appointed representative who is responsible for disaster preparedness and response at an event site.
    c. The lead individual responsible for evaluation, mitigation, and remediation after an event.
    d. An OSHA-certified individual who is ultimately responsible for properly allocating all appropriate resources in response to an event.

70. Which of the following depicts an accurate statement regarding the natures of "inspections" or "audits"?
    a. An audit generally requires the examination of every document associated with a certain topic.
    b. An inspection is typically broad in scope.
    c. The general objective of an inspection is to ensure that a specific task list has been completed at a predefined frequency.
    d. An audit usually focuses its review on a particular, single facet of a management system.

71. Which of the following is typically NOT a tool used by accident investigators at a worksite?
    a. Soil sampling equipment.
    b. Rope or security tape
    c. Photography or videography
    d. Electronic surveying equipment

72. One of the most important facets for re-planning and re-assessing project costs is that if more than _____ has passed, inflationary and contingency factors should be accounted for.
    a. Two years
    b. One year
    c. Six months
    d. Three months

73. Which of the following is regarded as a structured "brainstorming" technique to identify hazards of a systemic nature that can potentially lead to an incident or nonconforming product?

   a. A failure modes and effects analysis
   b. A hazard and operability analysis
   c. An event-tree analysis
   d. A workplace hazards assessment

74. What is conventionally considered the first step in conducting a job safety analysis?

   a. Scoping out potential hazards associated with a particular function or task
   b. Verification that worker competence is commensurate with associated responsibilities of the function or task
   c. Spending time observing workers performing the function or task, and developing a list of involved actions
   d. Review of prior-incident data historically associated with a particular function or task

75. OSHA's Voluntary Protection Program (VPP) is a program that requires a company to:

   a. Implement an integrated safety management program in its corporate protocol
   b. Have zero (0) recordable incidents over a period of two years to attain "Star" status
   c. Systematically improve its health and safety management system in a partnership between the employer and workforce
   d. Proactively utilize human performance initiatives and corrective action metrics

76. Which of the following is NOT a common initiator for gas leaks?

   a. Overpressurization
   b. Excessively cold temperatures
   c. Dirt contamination
   d. Filter cartridge mishandling

77. Structural failures are often attributed to changes in materials over a period of time. Such changes can affect a material's strength and ductility, thus eventually leading to a failure. Which of the following potential influences is typically NOT an initiator of such destructive changes?

   a. Stress
   b. Creep
   c. Shear
   d. Corrosion

78. Which of the following is a hazard associated with the handling of high-pressure fluids?

   a. Hydraulic neuropraxia
   b. Injection trauma
   c. Nitrogen-bubble formations (the "bends")
   d. Pascal's syndrome

79. Installed ramps must have a slope of less than ____ degrees for the general public and less than ____ degrees for handicapped access.
    a. 21, 16
    b. 18, 14
    c. 15, 11
    d. 12, 7

80. Freeze plugs are used with water and water-based liquids to control hydraulic pressure. They operate under the principal of the subject liquid expanding as it approaches its freezing point. As the liquid gets colder and expands, the freeze plug allows it to _____.
    a. Condense
    b. Vaporize
    c. Drain
    d. Sublimate

81. Which of the following is NOT a typically implemented control for preventing a materials-handling accident?
    a. Creating linear, narrow passageways or lanes for movement of materials outside of work areas
    b. Training workers to properly use hand signals
    c. Training workers on proper heavy-lifting techniques
    d. Implementing traffic controls for lift areas.

82. In regard to soil or other earthen material, the "angle of repose" is best defined as:
    a. The angle at which load-transferring underground columns are grounded relative to the trench angle within which they are emplaced
    b. The natural angle that soil forms when it is piled up or when it collapses
    c. The angular difference between an installed footing and its collocated boring
    d. The angular difference between an installed pile and its collocated footing

83. What is the "range" of the following set of values: 12, 3, 19, 8, 37, 5, 26, 44, 19?
    a. 9
    b. 41
    c. 19
    d. 19.2

84. A "leading indicator" is best defined as which of the following?
    a. Metrics that have happened or have been measured after a workplace injury has occurred.
    b. The most likely action or event responsible for initiating an occurrence or accident.
    c. An objective measure that is used to assess proactive actions taken to improve organizational performance.
    d. A determined "root cause" in a root-cause analysis.

85. Which of the following is NOT considered a primary objective of workers' compensation laws?

   a. To save workers the time, trouble, and expense of litigation
   b. To keep injured workers from turning to charities
   c. To encourage employers to develop procedures that prevent and reduce accidents
   d. To institute appropriate accident investigation protocols in order to determine individual culpability

86. What is the most appropriate response for a significant "near miss" that occurs on a worksite?

   a. Responding to it as if it were an injury incident, by completing a root-cause analysis and developing action plans to prevent potential recurrences.
   b. Interviewing involved personnel and management to acquire all involved details, and then issuing a "lessons learned."
   c. Filing an incident report and submit it to corporate's senior certified safety professional or safety and health manager.
   d. Directing the subject employee(s) to a medical professional for a screening evaluation to ensure no injuries were actually sustained.

87. Routinely scheduled inspections are an integral part of an effective occupational health and safety system. As such, the frequency of inspections must be aligned with the _____ posed by an operation, in conjunction with _____.

   a. OSHA requirements / 29 CFR 1910.69
   b. degree of risk / regulatory requirements
   c. assessment schedules / management directives
   d. auditing requirements / OSHA mandates

88. Which of the following would NOT be considered a "noneconomic" cost associated with an occupational injury?

   a. Morale of the injured worker as well as that of co-workers
   b. Employee absenteeism due to the workplace being considered hazardous
   c. Administrative or procedural upgrades to deficient work processes
   d. Poor productivity

89. Which of the following procedures is NOT typically regarded as a safeguard against machine and tool accidents?

   a. Ensuring a regular inspection and maintenance schedule
   b. Ensuring proper lockout and tagout protocol while machines or tools are being used
   c. Ensuring operators have no loose clothing, jewelry, or hair while machines or tools are being used
   d. Ensuring proper functioning of guards and guard devices

90. Statistical "variance" provides quantitative information about the degree to which a set of data values is spread apart, and also conveys how close an individual value is to the average relative to the other values in the set. "Variance" is thus calculated as which of the following:

   a. (standard deviation)² ÷ (mean)
   b. (mean)² ÷ (standard deviation)
   c. (standard deviation) ÷ (mean)²
   d. (standard deviation)²

91. Which of the following is true regarding environmental hazards?

   a. Effects often cannot be observed directly
   b. Effects are generally cumulative and typically only appear over time
   c. Radiation can be an environmental hazard
   d. All of the above

92. Which of the following is NOT a typical mode or type of respiratory protection used in the workplace?

   a. Free-oxygen respirators
   b. Air-purifying respirators
   c. Supplied-air respirators
   d. Self-contained respirators

93. Which of the following should NOT be deemed as a relevant factor when evaluating potential chemical hazards?

   a. Compounds that aren't normally dangerous on their own may become dangerous when mixed with other compounds.
   b. Diatomic gases can be more dangerous and volatile than monoatomic gases, since they exist at twice the concentration level of the latter.
   c. Substances that are known hazards may not be dangerous at lower concentrations
   d. Chemicals that "normally" aren't dangerous may become dangerous when used in certain ways.

94. Which of the following is true of personal protective equipment (PPE)?

   a. PPE is capable of removing hazards in certain scenarios.
   b. PPE is typically not considered within the construct of a safety plan.
   c. A HAZMAT suit does not fall under the official category of PPE.
   d. PPE should never be seen or used as a primary means for controlling hazards.

95. Which of the following is NOT a typical route through which a hazardous substance can enter the body?

   a. Ingestion
   b. Inhalation
   c. Diffusion
   d. Absorption

96. Which of the following substances would be appropriate for extinguishing a fire caused by flammable liquids?
    a. Halon
    b. Water
    c. Hydrogen peroxide
    d. Dry powder

97. The domain of "worksite sanitation" does NOT typically include which of the following:
    a. Safe (potable) drinking water
    b. Garbage disposal
    c. Clean areas to prepare and eat food
    d. Insect and rodent control

98. Which of the following is NOT considered a "repetitive strain injury"?
    a. Bursitis
    b. Carpal tunnel syndrome
    c. Stenosis
    d. Tendonitis

99. Which of the following is NOT considered an accurate statement regarding the use of manlifts?
    a. An operator or user can easily fall off a manlift platform.
    b. An operator or user can easily become caught as the manlift moves through an opening in the floor.
    c. An operator or user should avoid using a manlift if they suffer from Reynaud's syndrome.
    d. Per OSHA requirements, only trained personnel are permitted to operate manlifts.

100. When developing a maintenance checklist, which of the following should be consulted most heavily in determining what items need to be checked and at what frequency?
    a. ANSI/ASTM standards
    b. Manufacturer specifications and product literature
    c. System procedures
    d. ASME recommended maintenance schedules

101. Which of the following is NOT typically employed for controlling exposures to chemical hazards?
    a. Personal protective equipment
    b. Work modifications
    c. Toxicity reports
    d. Personal hygiene

102. Which of the following is true for "relief valves"?
    a. They are only used for gases and vapors.
    b. They are usually mounted between flanges in a vent pipe.
    c. The amount that they open depends on the quantity of underpressurization in a system.
    d. They close automatically when system pressure returns to normal.

103. Which of the following is NOT a typical hazard associated with the use of hand vehicles such as dollies, carts, wheelbarrows, and hand trucks?

a. Running into walls, equipment, or other people
b. Lack of visibility
c. Skill of the worker
d. Shifting, tipping, or falling loads

104. A "lagging" indicator is best defined or exemplified by which of the following?

a. A metric or occurrence that has transpired after a workplace injury has occurred.
b. A pre-employment physical that warned of higher incident risk to a worker due to a health condition.
c. The extent (or lack) of training provided to an injured worker after an incident.
d. An event that coincidentally occurs soon after the performance of a related risk assessment.

105. Which of the following is true of hand-tool safeguarding?

a. Shorter handles on axes and hatchets can enhance user safety.
b. Bent handles on various hand tools can help prevent tissue compression disorders.
c. Metallic handles with low coefficients of friction can help enhance a user's grip.
d. Non-powered hand tools must employ a lockout or tagout procedure similar to that for power tools.

106. Which of the following is NOT a typically employed safety measure associated with ladder use?

a. Ensuring that ladder rungs are slip-resistant
b. Placing ladders far enough from a wall so that toes can be comfortably emplaced on the rungs
c. The potential evaluation of nearby electrical conductors
d. Anchoring or tying ladders to a support structure

107. Which of the following is generally true regarding the use of hearing-protection PPE?

a. Earplugs or muffs can be used, but should not be combined due to potentially over-deafening oneself on the job.
b. Muffs are typically more effective against hearing loss than earplugs.
c. Per OSHA guidelines, only earplugs or muffs that are NIOSH-certified should be used in the workplace.
d. Hearing tests or audiograms should be scheduled at regular intervals for employees who work in environments with unprotected ambient eight-hour TWAs of greater than 35 dB.

108. Which of the following is NOT a type of *ionizing* radiation?

a. Gamma rays
b. Alpha particles
c. X-rays
d. Microwaves

109. Which of the following is NOT a typically employed safety measure associated with scaffold use?

a. Selecting a scaffold that is aptly load rated
b. Regularly checking outrigger beams for damage
c. Regularly checking planks, bolts, ropes, bracing, and clamps for damage
d. Using safety belts or lanyards when working on scaffolding at a height of 15 feet or more above ground level.

110. Which of the following actions is considered beneficial to reducing the risk of tripping hazards?

a. Avoiding changes in elevation that consist of one or two steps
b. Having only small changes in elevation between adjoining floor surfaces
c. Only placing electrical cords over walking paths during less-frequented employee hours
d. Wearing steel-toed shoes

111. At a construction site, which of the following *ionizing* radiation sources could a worker most likely be exposed to?

a. Gamma radiation from survey-instrumentation calibration sources
b. Radium and radon within earthen materials
c. Radiofrequency emitters
d. Neutron radiation from spontaneous fission within earthen uranium ore traces

112. To effectively follow up on and correct audit findings, subject items should be appropriately organized into which of the following categorization groupings?

a. Violations - infractions - notes
b. Deficits - actions - recommendations
c. Programmatic fault - marginal compliance - enhancement actions
d. Nonconformances - areas of concern - opportunities for improvement

113. The purpose of tracking incidents and collecting data is to use that data as a _____ tool that provides feedback on performance and resultantly guides active response.

a. line-management
b. human-performance
c. continuous-improvement
d. lessons-learned

114. A well-stocked first-aid kit would ordinarily NOT include the following:

a. Disposable sterile gloves
b. Mouthpieces
c. Syringes
d. Scissors

115. Which of the following is NOT an accurate categorization for explosives, as per U.S. Department of Transportation regulations?

a. Class "D" – includes fireworks and other minimally hazardous explosives
b. Class "C" – includes items that contain restricted amounts of Class A or Class B explosives
c. Class "B" – Explosives that function by rapid combustion instead of by detonation
d. Class "A" – Explosives that function by detonation

116. According to the "Human Factors Theory of Accident Causation," accidents are primarily caused by human error resulting from which of the following groupings?

a. Shortcutting – procedural errors – lack of investment
b. Worker apathy – management apathy – corporate apathy
c. Overload – inappropriate response – inappropriate activities
d. Anxiety – morale – ethics

117. Besides keeping body parts and clothing away from hazardous machine components, what is the other PRIMARY function of machine guards?

a. To capture and enclose dust
b. To prevent flying debris from striking people
c. To reduce machine noise
d. To provide a grounded electrical barrier between machine and worker

118. Any safety program(s) that reduce(s) _____ accidents will likewise reduce workers' compensation claims.

a. worker liability for
b. the number of fatal
c. the frequency and severity of
d. the high-deductible cost of

119. Risk is best defined as the multiplicative product of which of the following?

a. Cost-basis × fatalities
b. Occurrence × cost-basis
c. Remedial cost × frequency
d. Probability × consequence

120. Which of the following is NOT true regarding a standard normal distribution of data?

a. 68 percent of the values fall within one standard deviation of the mean of the data set
b. 85 percent of the values fall within two standard deviations of the mean of the data set
c. 99.7 percent of the values fall within three standard deviations of the mean of the data set
d. A normal distribution is typically depicted by a bell-shaped curve

121. "Book value" is best defined as which of the following?

a. Assets minus liabilities
b. Market value minus liabilities
c. Purchase price minus depreciation
d. Revenue plus margin

122. What are the four crucial elements for implementing a successful personal-protective-equipment program?

a. Selection, management, use, maintenance
b. Identification, procurement, review, hygiene
c. Need, use, mitigation, stewardship
d. Designation, protocol, ownership, compliance

123. Which of the following best defines the term "latency period"?

a. The amount of time required for an exposed worker to effectively eliminate an incurred toxin from the body.
b. The six-week period during the first trimester of pregnancy in which an unborn fetus can be the most susceptible to injury due to a hazardous exposure incurred by the mother.
c. The amount of time between a hazardous exposure and observable effects due to that exposure.
d. The amount of time a worker displays symptoms or sickness resulting from a hazardous exposure.

124. What is the primary difference between an "acute" and "chronic" exposure?

a. Acute refers to only one exposure, whereas chronic is always depicted by many exposures.
b. Acute exposure cases are usually never as severe as chronic exposure cases.
c. Workers' compensation is usually only awarded based on chronic exposures or incidents.
d. An acute exposure is always stochastic in nature, whereas chronic exposures are always somatic.

125. An "asphyxiant" is best defined as which of the following?

a. A material that causes irritation or coughing, but typically has no long-term dangerous effects on the body.
b. A material that displaces oxygen, thus interfering with breathing and oxygen-transport in the blood.
c. A substance that can potentially change the genetic structure of an animal or human.
d. A form of compressed nitrogen that can cause the lungs to bleed if inhaled.

126. What is the primary difference between "local effects" and "systemic effects" to the human body from a hazardous exposure?

a. Local effects often affect biological functions whereas systemic effects rarely do so.
b. Local effects and systemic effects are virtually the same entity and typically affect the body in a similar fashion.
c. Local effects frequently cause degrees of organ damage whereas systemic effects usually result in acute death.
d. Local effects typically affect the eyes, skin, or respiratory system, whereas systemic effects usually impact organs and body functions.

127. Which of the following is typically NOT true of burns?

a. A first-degree burn is superficial, with some reddening and pain.
b. A second-degree burn is deep, with blisters and pain, and usually takes a few weeks to heal.
c. A third-degree burn actually destroys the skin.
d. A third-degree burn usually has the most pain of the three types.

128. The most common type of workplace injury is:

a. Cuts
b. Sprains and strains
c. Bruises
d. Fractures

129. What is the most frequently injured part of the body in the workplace?
   a. Back
   b. Legs
   c. Arms
   d. Hands and fingers

130. Which of the following is NOT usually a common cause of workplace injuries or incidents?
   a. Overexertion
   b. Being struck by an object
   c. Slips, trips, and falls
   d. Overexposure to hazardous substances

131. Which of the following is NOT usually considered one of the most common causes of workplace accidents?
   a. Morale
   b. Willful unsafe practices
   c. Worker mismatch or overload
   d. Equipment traps

132. Which of the following is NOT usually considered an engineering control for protecting workers from hazardous exposures in the workplace?
   a. Substitution
   b. Isolation
   c. Remediation
   d. Ventilation

133. What is the primary purpose of a lanyard in a fall-protection system?
   a. Lanyards connect fall arrestors to lifelines.
   b. Lanyards connect safety harnesses to anchoring points.
   c. Lanyards connect safety belts to lifelines.
   d. Lanyards connect fall arrestors to safety harnesses.

134. In a construction area, "rigging" apparatuses are best defined as which of the following?
   a. Ropes, chains, or slings affixed between a lifting device and a load
   b. Hand-operated derricks or winches
   c. Cables affixed between the jib and trolley of a crane.
   d. Aerial baskets

135. Which of the following is NOT considered typical personal protective equipment?
   a. UV-rated sunglasses
   b. Rainwear
   c. High-visibility clothing and vests
   d. Coats or smocks

136. On average, as many as _____ of workplace injuries involving lost time are related to hand tools.
   a. 5 percent
   b. 8 percent
   c. 13 percent
   d. 17 percent

137. A "winch" is a type of:
   a. Gear
   b. Small crane
   c. Rigging
   d. Lifting device

138. A two-gallon container of methyl ethyl ketone (MEK) is found in a storage cabinet at a construction site. MEK has a flash point of 16°F and a boiling point of 175°F. What class of flammable liquid is MEK?
   a. IA
   b. IIA
   c. IB
   d. IIB

139. A job-hazard analysis is performed for an upcoming construction task in a normal-oxygen environment that may likely involve high levels of lead and beryllium dust. Which type of respiratory protection would be deemed most appropriate for this activity?
   a. A portable supplied-air system with blower and exhaust
   b. A portable vacuum hood
   c. A dust mask with an efficiency greater than 99.5 percent
   d. An air-purifying, full-face respirator

140. A respirator is equipped with a canister that is specifically designed for filtering ammonia gas. Per OSHA guidance, which of the following colors should the canister portray?
   a. Yellow
   b. Green
   c. Magenta
   d. White

141. Which of the following logs does NOT typically fall under the realm of "design documentation" for a construction project?
   a. Purchase order logs
   b. Shop drawing logs
   c. Request-for-information logs
   d. Submittal logs

142. Which of the following is typically NOT beneficial for enhancing construction ergonomics?
   a. Adjustable scaffolding
   b. Reduced vibration power tools
   c. Dolly aligners
   d. Motorized concrete screeds

143. Construction contractors should generate or maintain a(n) _____ that acknowledges: (1) daily weather conditions, (2) on-site employees and subcontractors, (3) deliveries of critical construction materials, (4) visits by third parties (e.g., project architects, owners, or engineers), (5) discoveries of previously unknown site conditions, discrepancies in plans, or conflicts, (6) important conversations that are held, and (7) any other noteworthy events.

    a. weekly report
    b. daily diary
    c. audit-traceability matrix
    d. Gantt logbook

144. What are the two primary U.S.-national accrediting agencies responsible for overseeing organizations that offer crane operator certification programs?

    a. OSHA and ANSI
    b. ASTI and AICM
    c. ANSI and NCCA
    d. COCA and OSHA

145. Which of the following should NOT be a metric used to determine whether worker "competency is commensurate with responsibility"?

    a. Education level
    b. Demographics
    c. Work experience
    d. Language proficiency

146. Which specific OSHA regulation addresses "Construction Worker Safety Training and Education"?

    a. 29 CFR 1926.21
    b. 29 CFR 1921.35, Appendix A
    c. 29 CFR 1916.48
    d. 29 CFR 1910.727, Subparts B & C

147. Which of the following is NOT designated as a prime carcinogen per OSHA 29 CFR 1910.1003?

    a. Benzidine
    b. Ethyleneimine
    c. 4-nitrobiphenyl
    d. Ammonium hydroxide

148. Which of the following individuals would ideally be the most qualified to measure a variety of hazards in the workplace and resultantly provide recommendations to mitigate such hazards?

    a. A certified health physicist (CHP)
    b. A certified industrial hygienist (CIH)
    c. A certified associate safety professional (ASP)
    d. An NSPE-registered industrial (professional) engineer

149. Rubber-tired self-propelled scrapers, rubber-tired front-end loaders, rubber-tired dozers, wheel-type agricultural and industrial tractors, crawler tractors, crawler-type loaders, and motor graders that are used in construction work shall all be equipped with _____.

   a. transmission interlocks
   b. compartmentalized ballast
   c. an independent emergency cutoff switch
   d. rollover-protective structures

150. The BCSP requirements for CHST recertification are the reporting of activities every ____ years, and a minimum of ____ recertification points.

   a. three, 15
   b. three, 25
   c. five, 20
   d. five, 35

151. General construction worksite (plant) areas and shops shall be illuminated at an intensity of no less than:

   a. 5 foot-candles
   b. 10 foot-candles
   c. 15 foot-candles
   d. 20 foot-candles

152. Which of the following tools or resources is NOT specifically ascribed to within the realm of chemical hazard communication protocol, as per OSHA 1910.1200?

   a. The U.N. Globally Harmonized System of Classification and Labeling of Chemicals (GHS)
   b. (Material) safety data sheets
   c. NIOSH's REL compendium
   d. Guidance per the U.S. EPA's Toxic Substances Control Act

153. Safety-toe footwear for construction employees shall meet the requirements and specifications stated within:

   a. American National Standard for Men's Safety-Toe Footwear, Z41.1-1967
   b. OSHA Standards 29 CFR 1910.45 and 29 CFR 1926.211
   c. NIST Standard for Safety Shoes and Safety Boots - N1412.91
   d. ISO Personal Protective Equipment Footwear Standard – ISO 11051A

154. Protective certified headwear (hardhats, helmets) must be worn in any construction area that entails hazards from direct impacts, falling objects, flying objects, burns, and _____.

   a. Potential vehicle use
   b. Slippery surfaces
   c. Falls
   d. Electrical shock

155. Construction employees working over or near water, where the danger of drowning exists, shall be provided with:

    a. Nearby ring buoys with at least 30 feet of line
    b. ANSI-rated flotation devices per Z45.2-1998
    c. U.S. Coast Guard-approved life jackets or buoyant work vests
    d. At least a two-hour-equivalent water safety training course, prior to commencement of work

156. A "Type-A" rated fire extinguisher is effective against which type of fire?

    a. Flammable liquids
    b. Trash, wood, or paper
    c. Electrical equipment
    d. All of the above

157. Per OSHA 29 CFR 1926.404 states that: All __-volt single-phase _____ receptacle outlets on construction sites, which are not a part of the permanent wiring of a building or structure and which are in use by employees, shall have approved ground-fault circuit interrupters for personnel protection

    a. 110, 10- and 15-ampere.
    b. 120, 20- and 30 ampere.
    c. 120, 15- and, 20 ampere.
    d. 110, 20+ ampere.

158. Per OSHA 29 CFR 1926.441, which of the following is NOT stipulated in regard to proper battery storage and charging protocol?

    a. Facilities for quick drenching of the eyes and body shall be provided within 100 feet (30.48 m) of battery handling areas.
    b. Unsealed batteries shall be located in enclosures with outside vents.
    c. Unsealed batteries shall be arranged so as to prevent the escape of fumes, gases, or electrolyte spray.
    d. When batteries are being charged, vent caps must be kept in place.

159. Construction worksite fall-protection training MUST include or acknowledge which of the following?

    a. Limitations on the use of personal protective equipment during the performance of low-sloped roofing work
    b. How to report perceived deficiencies in safety-monitoring systems
    c. The use and operation of guardrail systems and controlled access zones
    d. Procedures for conducting audits of fall-protection systems

160. Per OSHA 29 CFR 1926.555, which of the following is NOT a mandated requirement for conveyors used at a construction worksite?

    a. Conveyor systems shall be equipped with an audible warning signal to be sounded immediately before starting up the conveyor.
    b. If the conveyor operator station is at a remote point, a stop switch for the conveyor motor shall not be provided at the motor location.
    c. Emergency-stop switches shall be arranged so that the conveyor cannot be started again until the actuating stop switch has been reset to the "on" position.
    d. All conveyors shall meet the applicable requirements for design, construction, inspection, testing, maintenance, and operation, as prescribed in the ANSI B20.1-1957.

161. Employees engaged in site-clearing activities shall be protected from the hazards of _____.

   a. Irritant and toxic plants
   b. Dust inhalation (pica)
   c. Inadvertent non-potable water consumption used for dust suppression
   d. Carbon monoxide overexposure

162. Which of the following is NOT a typical "shoring" technique used for trenching and excavating?

   a. Aluminum hydraulic shoring
   b. Timber shoring
   c. Pneumatic shoring
   d. Base-mounted shoring

163. Which of the following construction tools or machines is NOT considered "high-risk"?

   a. Circular saw
   b. Nail gun
   c. Drill
   d. Conveyor

164. Which of the following is NOT a type of concrete or concrete-related operation used in the construction industry?

   a. Shotcrete
   b. Precast concrete
   c. Fiber-reinforced concrete
   d. Rubber cement

165. During demolition preparation activities, if any hazardous substance(s) is/are apparent or suspected, _____ shall be conducted, and the hazard is to be eliminated before demolition is started.

   a. a hazards assessment
   b. testing and purging
   c. a HAZMAT audit
   d. a JSA

166. Per OSHA 1926.961, which of the following is NOT considered typical standard operating procedure for the re-energization of lines and equipment?

   a. Ensuring that no one initiates action to reenergize lines or equipment at a point of disconnection until all subject components are verified as grounded.
   b. Ensuring that all crews working on lines or equipment release their clearances.
   c. Ensuring that all employees are physically clear of lines and equipment.
   d. Ensuring that all protective tags are removed from the point of disconnection.

167. Which of the following is NOT a safety requirement for temporary stairways installed for construction activities?

a. They shall have landings of not less than 30 inches (76 cm) in the direction of travel and must extend at least 22 inches (56 cm) in width at every 12 feet (3.7 m) or less of vertical rise.
b. Stairs must be installed between 20 and 55 degrees from horizontal.
c. Variations in riser height or tread depth shall not be over 1/4 inch (0.6 cm) in any stairway system.
d. Where doors or gates open directly on a stairway, a platform shall be provided, and the swing of the door shall not reduce the effective width of the platform to less than 20 inches (51 cm).

168. The term "fit-testing" is commonly used when referring to what type of personal protective equipment?

a. Full-body suits
b. Respirators
c. Fall-arresting harnesses
d. Cardiopulmonary pulse monitors

169. Whenever there is a concern as to safety, crane and derrick operators must have the authority to _____ until a qualified person has determined that safety has been resoundingly assured.

a. evacuate a work area
b. preliminarily examine such a concern
c. file a corrective-action request
d. stop and refuse to handle loads

170. Which of the following is a requirement for a masonry-project "limited access zone"?

a. The zone shall be equal to the height of the wall to be (re)constructed plus two feet, and shall run the entire length of the wall.
b. The zone shall be equal to the height of the wall to be (re)constructed plus two feet, and shall run 1.1 times the entire length of the wall.
c. The zone shall be equal to the height of the wall to be (re)constructed plus four feet, and shall run the entire length of the wall.
d. The zone shall be equal to the height of the wall to be (re)constructed plus four feet, and shall run 1.5 times the entire length of the wall.

171. A "surcharge load" is defined as a material or physical mechanism that can increase the likelihood of a trench cave-in or collapse. Which of the following would NOT be considered a type of surcharge load?

a. Heavy equipment situated or stored close to the trench
b. Spoil piles amassed close to the trench
c. Workmen standing near the trench
d. Workmen working inside the trench

172. Which of the following types of soil does NOT have a low compressive strength?

a. Sand
b. Clay loam
c. Gravel
d. Unstable rock

173. What is the most frequently occurring injury resulting from electric shock?

a. Electrocution
b. Acute fibrillation arrhythmia
c. Head impact injuries due to loss of consciousness
d. Burns

174. Which of the following measures or techniques would NOT be considered effective against controlling static electricity?

a. Humidification
b. Incorporation of additives to the air
c. Ionic dampening
d. Bonding or grounding

175. A steel-erection contractor shall not erect steel unless it has received written notification that the concrete in the footings, piers, and walls, or the mortar in the masonry piers and walls, has attained, on the basis of an appropriate _____ of field-cured samples, either ____ percent of the intended minimum compressive design strength or sufficient strength to support the loads imposed during steel erection.

a. ASTM standard test method, 75
b. ANSI standard test method, 85
c. ASME standard test method, 90
d. ISO standard test method, 50

176. To ensure scaffold safety, which of the following is a minimum requirement for an employee to be on a moving scaffold?

a. An outrigger is installed on at least one side of the scaffold
b. The scaffold height-to-base ratio is 2:1
c. The workers are located at scaffold positions outside the wheels
d. The scaffold platforms overlap at least six inches over their supports

177. The only time a body belt may be used where there may be a fall hazard is when a construction worker is using a(n) _____.

a. safety sling
b. arrestor
c. positioning device
d. braking mechanism

178. A "warning line system" is conventionally defined as which of the following?

a. An interlock system that prevents the decoupling of a lanyard from a snap hook
b. A series of concentric floor demarcations that depict incrementally increasing radiation levels within a radiation area
c. An erected barrier on a rooftop to warn workers they are approaching an unprotected side or edge
d. A pressure-differential gauge on a respirator system that indicates it is time for a filter-cartridge replacement

179. Potentially fatal "orthostatic incompetence" is defined as which of the following?
   a. The experienced medical effects of being immobilized or suspended in a vertical position
   b. Being forced to land head first into a safety net given the angle of a fall
   c. The bursting of a full bladder due to a fall impact
   d. A sudden drop in blood pressure due to a rapid increase in elevation or altitude

180. Which of the following is NOT a typical hazard associated with welding, cutting, and brazing operations?
   a. Air contaminants
   b. Potential eye damage from UV light
   c. Asphyxiation
   d. Explosions

181. Which of the following software tools would be most appropriate for safety-program recordkeeping and audit tracking?
   a. Project management software
   b. Database software
   c. Presentation software
   d. Spreadsheet software

182. Security and privacy of worker personal health records is covered by law under which of the following?
   a. OSHA
   b. HIPAA
   c. AMA
   d. AACN

183. Which of the following substances is commonly used in oxygen-fuel welding and cutting?
   a. Methane
   b. Polybutylene
   c. Hydrazine
   d. Acetylene

184. Which of the following is NOT a "common" hazardous occurrence resulting from the misuse or poor maintenance of a hand tool?
   a. Eye injuries caused by using a screwdriver as a chisel
   b. Pieces of wooden handles on hammers or axes breaking off and striking an individual
   c. Stripping a pliers' set-screw while in use
   d. Springing a wrench's jaws open while in use

185. "White finger" in the construction workplace is typically caused by which of the following?
   a. The continuous use of vibrating hand-held machinery
   b. The continuous action of hammering
   c. Keeping fingers locked in the same position for an extended period of time
   d. The reoccurrence of calluses and blisters in the same location over an extended period of time, ultimately resulting in a sense of numbness in that area

186. What is the MOST important safety rule to always remember when using a portable circular saw?

    a. Always use a sharp blade
    b. Never hold or balance a piece of material in one hand while trying to operate the saw with the other
    c. Always wear a face shield in tandem with safety goggles
    d. Always ensure that the saw's switch actuates properly

187. Lifting frequency is the average number of lifts performed over a _____ minute period.

    a. 60
    b. 45
    c. 30
    d. 15

188. Which of the following is NOT an example of a compressed-air (pneumatic) tool used on a construction site?

    a. Nail gun
    b. Chipper
    c. Brazing torch
    d. Sander

189. Which of the following is TRUE regarding OSHA inspections?

    a. The employer may or may not be advised by the compliance officer of the reason for the inspection.
    b. The employer must accompany the compliance officer on the inspection.
    c. The compliance officer must show two separate official forms of identification upon an inspection.
    d. The employer must agree to submit to an OSHA audit within a six-month period if more than two adverse conditions are documented by the compliance officer.

190. A construction company had three (3) recordable injuries, with one of them resulting in 65 days of lost time. The total number of hours worked was 248,620. What is the calculated "severity rate" associated with these statistics?

    a. 52.29
    b. 0.019
    c. 0.046
    d. 21.74

191. The "tangent" of ∠ABC is equal to:

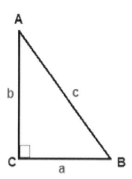

   a. a/c
   b. b/a
   c. a/b
   d. b/c

192. A strong acid typically has a pH around ___, and a strong base typically has a pH around ___.

   a. 7, 14
   b. 14, 0
   c. 1, 14
   d. 0, 7

193. Which of the following agents or substances in the workplace is NOT likely to result in disease or a disorder if repeated exposures occur?

   a. Beryllium
   b. Cadmium
   c. Argon
   d. Silica

194. Which of the following is NOT a type of ventilation-measurement equipment?

   a. Pitot tubes
   b. Damper hydrocouplings
   c. Rotating vane anemometers
   d. Thermal anemometers

195. Which of the following is NOT an official OSHA violation category?

   a. Willful violation
   b. Repeated violation
   c. Mitigation violation
   d. Serious violation

196. Per OSHA's audiometric testing standards, a baseline audiogram must be conducted within ___ month(s) of a first exposure greater than ___ dBA.

   a. 6, 85
   b. 2, 80
   c. 3, 75
   d. 1, 90

197. Which of the following is NOT a cold-stress-related condition, injury, or illness?

a. Trench foot
b. Chilblains
c. Hyperthermia
d. Frostbite

198. The term "TLV," in regard to hazardous chemical exposure, stands for which of the following?

a. Teratogen level variant
b. Threat level variable
c. Theoretical latency vector
d. Threshold limit value

199. A basic multimeter can be used to measure which of the following electrical parameters?

a. Voltage, current, and capacitance
b. Inductance, amplitude, and resistance
c. Voltage, impedance, and resistance
d. Voltage, current, and resistance

200. Which of the following is NOT a typical option used for the delivery of safety training?

a. Needs-analysis training
b. Instructor-led training
c. Self-paced training
d. Structured on-the-job training

Answer Key and Explanations

1. B: The end-result of a potential electrocution incident is generally determined by how much current passes through the body, the duration of time to which a person is exposed to the subject current, and the actual path the current takes.

2. C: An eight-hour time-weighted average of 85 dB is considered the threshold, per 29 CFR 1910.95, for which a Noise Hazard Control Program (and associated Hearing Conservation Program) must be set into place.

3. A: Per 29 CFR 1910.1001, an employer shall ensure that no worker under its employment is exposed to an airborne concentration of asbestos in excess of 0.1 fiber per cm^3 of air as an eight (8)-hour time-weighted average (TWA).

4. B: OSHA regulations 29 CFR 1910.25-1910.27 stipulate that no single ladder or individual ladder sections shall exceed 30 feet in length.

5. C: Certain attributes for hazardous waste are stipulated and codified within EPA regulations (40 CFR 261, Subpart C) for special designation. The four properties, in order, are ignitability (# D001), corrosivity (# D002), reactivity (# D003), and toxicity (#'s D004-D043). Mutagenicity is not designated as a codified attribute within 40 CFR 261, Subpart C.

6. B: Robust, form-fitting goggles that fully and tightly cover the eyes provide very good protection from airborne dust hazards, and can also provide a suitable optical safeguard against chemical splashes and airborne projectiles.

7. A: Insulating gloves are required to be periodically tested at minimum six-month intervals, with the associated testing procedures and results comprehensively annotated. As an extra precautionary measure against hand injuries, it is advisable that workers wear leather gloves in tandem with the insulating gloves to provide additional protection.

8. D: A job safety analysis (JSA) is likely the most suitable choice for a small-scale worksite because workers who are familiar with the site and its associated functions and tasks actually perform and render the subject analysis. In contrast, a failure modes and effects analysis is a systematic method for evaluating a process to identify where and how it might fail and for assessing the relative impacts of different failures; an event tree evaluates how single events may logically spawn subsequent events or consequences; and a root-cause Analysis is a method of problem solving that identifies underlying faults or deficiencies that likely initiated an event.

9. D: When 25 or more workers are underground at one time, two separate rescue teams must always be available to aid and assist in case of an emergency; one of the teams must be within 30 minutes of traveling time to the underground location, and the other must be within two hours of traveling time. If fewer than 25 workers are underground at one time, only one rescue team needs to be available, and they must be located within 30 minutes of traveling time to the subject location.

10. A: Clay is a Type-A material and requires that a maximum 53-degree slope not be exceeded, with an associated height-to-depth ratio of ¾ to 1. Thus, a five-foot trench requires, at most, a 3.75-foot angle, or, 53 degrees. Any angle greater than this can result in a cave-in. Type B is gravel and silt, which require that a 45-degree slope not be exceeded. Type C is sand, which requires that a 34-degree slope not be exceeded.

11. B: Lifting index (LI) measures the physical stress associated with lifting an object. As LI increases, the chance of injury likewise increases. LI is calculated by dividing the load weight by the recommended weight limit (RWL).

12. D: Flash point is defined as the lowest temperature at which a flammable liquid can form an ignitable mixture in air. If the source of ignition is ultimately removed, the vapor will likely cease burning.

13. C: To ensure a safe lift, a crane operator must be aware of whether a crane's outriggers are extended or retracted, the angle of the boom, and whether the crane's tires are properly inflated. In addition, the operator must ensure that the crane is level, that extended outriggers are supported by stable ground, what positions the boom will be in during the lift, and the gross weight of the load. All of this information can be used for consulting a load chart that will assist the operator in determining whether a given load is within the structural and stability limits of the crane. Although the width of the jib has no meaningful significance in such a determination, it is vital to be aware of the length of the jib.

14. B: A body harness consists of straps attached to other components of a personal fall system. In case of a fall, the straps primarily distribute the fall-arrest force over the chest, shoulders, waist, thighs, and pelvis. A properly designed (and worn) harness should yield minimal impacts to the lower-back area.

15. A: Recommended weight limit (RWL) is the weight that healthy workers could lift for up to eight hours without causing musculoskeletal injuries. It is calculated as the multiplicative product of load constant, horizontal multiplier, vertical multiplier, distance multiplier, asymmetric multiplier, frequency multiplier, and coupling multiplier.

16. D: The incident-injury ratio Heinrich developed is 300:29:1. This ratio demonstrates, statistically, that an attentive manager or foreman usually has many opportunities to improve a safety program before a serious accident occurs.

17. C: The three E's of safety are engineering, education, and enforcement. Engineering entails the use of safety processes and procedures; education emphasizes worker training and hazard identification; and enforcement translates to obligatory compliance with rules, laws, and regulations.

18. B: Total case incident rate (TCIR) is a health and safety metric that represents the number of OSHA-recordable injury cases in a year per hundred full-time employees. The metric is primarily used for comparison between entities in similar industries.

19. B: Falls at construction worksites usually represent the highest source of fatalities (about one-third). Other major sources include transportation-related fatalities (about 25 percent), contact with objects and equipment (about 20 percent), and exposure to harmful substances (about 15 percent).

20. A: Per OSHA regulation 29 CFR 1926.651(c)(2), a stairway, ladder, ramp, or other safe means of egress shall be located in trench excavations that are four feet (1.22 m) or more in depth so as to require no more than 25 feet (7.62 m) of lateral travel for employees.

21. B: OSHA regulations mandate that trenches five feet (1.5 meters) deep or greater require a protective system against cave-ins unless the excavation is made entirely in or of stable rock. Trenches 20 feet (6.1 meters) deep or greater require that the cave-in protection system be designed by a registered professional engineer or be based on tabulated data prepared or approved by a registered professional engineer.

22. D: Although any amount of electrical current over 10 milliamperes typically produces painful to severe shock, death usually does not occur at levels below 75 milliamperes.

23. D: Overhead power lines are usually not insulated, and thus provide a major electrical hazard if direct contact is established with any conductive material. Construction equipment such as backhoes, scaffolding, dump-truck beds, cranes, ladders, scissor lifts, and even aluminum paint rollers may inadvertently come into contact with such lines. Bulldozers are generally not tall enough to hit overhead power lines.

24. C: Workers below scaffolds are often struck by objects such as tools or materials that inadvertently fall from the scaffolds. To mitigate the potential of such objects striking workers, it is recommended that workers always wear hard hats; the area beneath scaffolds be barricaded to the highest extent practicable; panels or screens be used to contain and stabilize stacked objects; and canopies or nets be erected to catch or deflect falling objects. There are no prescribed regulations in existence that mandate a safety buffer margin or distance below a scaffold.

25. C: OSHA regulation 29 CFR 1926.502 stipulates that when a 200-pound load is applied in a downward direction to the top edge of a guardrail, it shall not deflect to a height any less than 39 inches above the walking or working level. Additional fall-protection measures must additionally be installed (such as mid-rails, screens, mesh, and intermediate vertical members) between the top edge of the guardrail system and the walking or working surface when there is no wall or parapet wall present at least 21 inches in height.

26. A: The three principal welding modes used in the construction industry are oxygen-fuel gas welding, arc welding, and resistance welding. Oxygen-fuel gas welding is a process that uses fuel gases to weld, by heating metals to a temperature that produces a shared pool of molten metal, which cools into a common metal. Arc welding is a fusion process wherein the merger of metals is achieved from the heat of an electrode and a bond formed between the elements. Resistance welding utilizes pressure and heat that is generated in the components to be welded by resistance to an electrical current.

27. C: The "ring test" evaluates the sound coming from a grinding wheel when lightly tapped with a nonmetallic material. An undamaged wheel emits a clear ringing tone. The ring test should not be performed on wheels that have a diameter of 10 cm or less, plugs and cones, mounted wheels, segment wheels, or inserted nut and projecting stud disc wheels.

28. B: Soil classifications are A, B, and C. Soil type A is cohesive soil with a high compressive strength, soil type B is cohesive soil with a moderate compressive strength, and soil type C is cohesive soil with a low compressive strength.

29. D: Electrical bonding is defined as the connection of two or more conductive objects with a conductor to prevent electrocution. Typical bonding connectors include copper bonding straps, solder lugs, pressure connectors or bolts, and star washers. A typical bonding conductor is a copper braid with pressure-type clamps.

30. C: Per OSHA regulation 29 CFR 1910.242, only compressed air of less than 30 psi may be used for cleaning operations.

31. A: Electrical grounding is the process of connecting one or more conductive objects directly to the earth using cold water pipes, building steel, or ground rods.

32. C: Per OSHA Regulation 29 CFR 1915.159, anchor points for personal fall arrests should be capable of holding at least 5,000 pounds of force per person.

33. D: Cranes, derricks, and other tall equipment or structures must be clear of power lines by at least 10 feet for voltages up to 50 kV and 10 feet plus 0.4 inch for each kV over 50.

34. B: Although <u>all</u> involved entities should take ownership and accountability for all facets of job site safety, it is ultimately the employer's responsibility to ensure that only well-maintained and operable hand and power tools are utilized at a job site.

35. C: During the course of construction, alteration, and repair efforts, all scrap lumber and associated debris must be regularly cleared from work areas, passageways, and stairs, and from around buildings or other structures. There is no prescribed time interval for this requirement, however (daily or otherwise). Safe housekeeping practices for disposing of all construction site refuse and scrap (including combustibles and hazardous wastes) are provided in 29 CFR 1926.25.

36. B: Under current U.S. laws for workers' compensation, there are four potential injury categories: "partial" – when the employee can still work, but is unable to perform all duties of the job due to the injury; "total" – when the employee is unable to work or perform substantial duties on the job; "temporary" – when the employee is expected to fully recover; and "permanent" – when the employee will suffer the effects of an injury for the rest of his or her life.

37. B: Heat stroke occurs when the body's temperature regulation system fails and sweating becomes inadequate or stops altogether. It is a medical emergency situation and without immediate treatment administered, brain damage or death is very likely to occur. Signs and symptoms of heat stroke include mental confusion, euphoria, redness in the face, chills, restlessness, irritability, disorientation, hot skin, cessation of sweating, erratic behavior, shivering, collapse, unconsciousness, convulsions, and a core body temperature which exceeds 104 degrees F. To prevent the occurrence of heat stroke, workers should steadily adapt themselves to their work environments, follow an appropriate work-rest cycle, drink plenty of fluids, and maintain a proper diet.

38. A: OSHA requires that engineering controls, personal protective equipment, and administrative controls be used in the workplace to prevent work-related injuries.

39. C: The ISO 14001 environmental management system is built on upon the plan-do-check-act model. The clauses of the standard require that a working organization establish an environmental policy, that they commit to compliance with all applicable environmental regulations, that they adequately communicate their environmental programs to employees, and that the organization periodically conducts audits to assess compliance levels.

40. C: A confined space, by definition, does not have a continuous-air-supply or ventilation system attached to its construct. Thus, workers and safety professionals must always be aware of the possibility that the subject area may be an oxygen-deficient environment and that a significant buildup of CO_2 over time is also a distinct possibility.

41. D: The only stated condition that would be indicative of a hazardous atmospheric environment is the presence of a flammable gas at a concentration of 25 percent of its lower flammability limit (LFL). Acceptable or normal atmospheric concentrations of carbon dioxide are typically less than

600 ppm, and atmospheric oxygen concentrations are considered normal within the range of 19.5-23.5 percent. An inert gas would only be dangerous if it existed in high enough concentration to displace breathable air.

42. B: Directly measuring work environment temperatures, as well as verifying the proper operation of applicable HVAC systems, are appropriate measures to invoke for determining the extent of heat or cold to which a worker may be subjected within an open or otherwise confined work area. In addition, various vital-sign measurements (pulse, blood pressure, electrolyte levels, etc.) may also be invoked as appropriate. A worker, however, should NEVER be asked to enter into a potentially adverse area without prior independent verification (through instrumentation or other noninvasive evaluation) that the area is suitable for entry.

43. D: Combustible (and flammable) liquids not stored within the confines of an outside storage area must be completely surrounded by a curb structure or enclosure that is at least six inches in height. Aspects of periodic filter change-outs, anchoring, and sun exposure may also hold applicability on a case-by-case basis with regard to a worksite's specific protocol for handling and storage of such materials.

44. A: Per OSHA 29 CFR 1904.6, the time requirement for retention of documents (e.g., the OSHA 300 Log) is designated to be five years following the end of the calendar year that the records cover.

45. A: The OSHAS 18000 series of occupational health and safety management system standards sets the basis for an effective system to control the health and safety of workers. The purpose of such a management system is to take a systemic approach to the manner in which an organization manages its health and safety protocol, with the overall goal being to proactively identify and respond to risk before an accident or incident occurs.

46. B: Per OSHA 29 CFR 1910.211(b), cut-off wheels are recommended only for use on specially designed and fully guarded machines. A wheel with a diameter in the range of 6 to 12 inches can cut a material with a maximum thickness of ¼ inch.

47. C: Per OSHA 29 CFR 1910.1025, the permissible exposure limit for lead is 50 micrograms per cubic meter (50 ug/m^3) of air, averaged over eight hours. This means that the employer shall ensure that no employee is exposed to lead concentrations greater than 50 micrograms per cubic meter of air over an eight-hour averaged period. If an employee is exposed to lead for more than eight hours in any work day, the permissible exposure limit, as a time-weighted average (TWA) for that day, shall be reduced per the following formula: permissible exposure limit = (400) ÷ (hours worked in the subject day).

48. B: Training employees on the OSHA Hazard Communication Standard is required whenever a new hazardous substance is potentially brought into the work environment, or in conjunction with an employee's initial assignment to a specific work area. It is recommended, however, that the subject material be easily available or extractable to all worksite employees on an as-needed basis.

49. A: There are no specific requirements stating that tags must be red or orange in color, must be attached in two separate places with a metal connecting device, or must display the term "Hazard" or "Electrical Hazard" on their face(s). Tags must, however, be normalized (standardized) according to size and shape. In addition, they must be strong enough and affixed well enough to prevent any inadvertent detachment, and must always identify the employee(s) who applied them.

50. C: The American National Standards Institute (ANSI) is the agency empowered with the ongoing responsibility of regularly setting standards for safety glasses and other safety-related eyewear.

51. A: Inadequate/substandard work procedures would most likely result from "substandard management/oversight practices" as an incident causal factor. The rendering or implementation of sound procedures and work practices is ultimately the responsibility of organizational management.

52. C: The CHST can best identify workplace hazards through the conduct of a rigorous worksite analysis. The employment of such tools as fault-tree analyses, FMEAs, and trend assessments are typically more invoked at the engineer and CSP levels.

53. A: A willful violation can result in a fine of up to $250,000 upon a criminal conviction, if an employee death occurs.

54. B: The constant for gravitational acceleration is 32.2 ft/sec². In metric terms, this equates to 9.8 m/sec². The constant depicts the acceleration on an object caused by the force of gravity. Neglecting friction such as air resistance, all small bodies accelerate in a gravitational field at the same rate relative to their center of mass. This holds true regardless of the masses or compositions of the bodies.

55. C: For the presented right triangle, to determine the length of the base "a," the Pythagorean Theorem may be invoked per the following:

$$a^2 + b^2 = c^2$$

$$a = \sqrt{c^2 - b^2} = \sqrt{169 - 144} = \sqrt{25} = 5$$

56. B: A worksite security evaluation is a typical constituent of a worksite analysis following a workplace violence incident or event.

57. D: Lost-time trend evaluations are not considered safety "interventions" because they do not (probabilistically) have a direct effect on potentially reducing the likelihood of an accident or event. Measures such as installing design or engineering upgrades, enhancing safety program participation, and conducting robust training programs can all have a measurable effect on lowering workplace incident risk.

58. C: Per OSHA 29 CFR 1910.23 (Guarding Floor and Wall Openings and Holes), a floor-hole is defined as a (measurable) hole with a range between 1 and 12 inches.

59. D: Per ANSI/ASSE A1264.2, a coefficient of friction close to zero ('0') means that a surface is regarded as extremely slippery.

60. B: Rubber-insulated gloves used in real-time electrical operations and construction work over 50 volts must meet the threshold requirements prescribed in ASTM D 120-09 (Standard Specification for Rubber Insulating Gloves) and must also be categorized as "Type-I."

61. C: Overloading, visibility issues, and tipping are some of the most common hazards associated with the operation of large construction vehicles such as bulldozers, backhoes, and forklifts. High carbon monoxide production should not be an issue as long as the construction area is outdoors or well ventilated.

62. D: Ring guards, enclosure guards, and interlocked guards are all examples of point-of-operation guarding. Ring guards typically enclose a rotating cutter, enclosure guards keep body parts or clothing from contacting a point of operation, and interlocked guards prevent a machine from operating when a section is open.

63. B: Safety nets, harnesses, and catch platforms may all be used at construction worksites as fall-limiting devices. Pitons and safety cables are typically used by advanced rock climbers.

64. A: Anti-fatigue mats are often employed in the workplace for reducing standing hazards such as back pain, sore feet, and varicose veins. The mats provide a cushion between a worker's feet and hard floor surfaces, especially concrete.

65. D: OSHA guidelines mandate that fall-protection equipment be implemented whenever a construction worker is at least 6 feet above ground level, or if an employee is on scaffolding at least 10 feet above ground level.

66. D: Pipes and bars should not be stored in racks facing main aisles, as they pose a hazard to passerby when removing supplies. Employing a stepping-back layout for materials stacked in several rows, and using crossties, retaining walls, racks, netting, and protective barriers are also considered robust controls for safe material storage.

67. B: During a rigging inspection, if the throat opening of a hook (which attaches to the rigging) has increased in excess of 15 percent of the original throat opening, it should be replaced. Other items to be evaluated during a rigging inspection include broken wires or fibers, corrosion, wear, kinking, crushing, cracks, chain stiffness, retainer positions, and fitting alignments.

68. B: During a shelter-in-place event, the primary intent is not to evacuate an area (although at a later point it may ultimately be decided that an evacuation may be necessary). Shelter-in-place procedures should include guidance on where occupants should gather, how to temporarily close off all exits and entrances, how to shut down ventilation systems and elevators, and how to inform occupants that an emergency has occurred.

69. A: An on-scene coordinator is generally defined as the individual who is in charge of coordinating various agencies and departments to ensure that all needs are responded to during and after an event.

70. C: An inspection generally refers to the checking of a list of items that are verifiable. It is typically narrow in scope and generally implemented to ensure that regulatory requirements are being met. Its overall objective is usually to ensure that a specific task list has been regularly completed at a predefined frequency. An audit, in contrast, generally refers to a review of an entire management system, with its overall objective usually being to examine a system designed to manage risk. It does not typically examine every document associated with a certain topic, but rather a representative sample.

71. A: Accident investigators typically manage and implement tools such as security rope, security tape, photographic and video equipment, tape measures, tape recorders, Geiger counters, sampling equipment, specimen jars, and electronic surveying equipment.

72. C: If more than six months have passed, increases for inflation or in vendor prices should be considered during a project re-plan or re-estimate effort. Contingency factors should also be considered to account for future uncertainties.

73. B: A hazard and operability analysis (HAZOP) is a structured technique to identify hazards of a more systemic or operational nature that can potentially lead to unexpected or undesired products or outcomes. It principally relies on a regimented "brainstorming" technique that guides analysts into considering all modes of potential deviation that can potentially lead to a failure.

74. C: A usual first step in the conduct of a job safety analysis is to spend time observing workers doing their jobs, and to develop specific lists of actions and processes associated with the subject tasks of the jobs in question.

75. C: The overarching goal of OSHA's Voluntary Protection Program (VPP) is to help systematically improve a company's health and safety management system by fostering a strong bond or partnership between the employer and workforce within the realm of health and safety. A company must submit an application to OSHA in order to be admitted to the program. OSHA awards "Star" and "Merit" program statuses to outperforming and overachieving entities.

76. D: Gas leaks usually occur due to residual dirt on valves, gaskets, and threads. Other potential factors that may lead to a leak include overpressurization, excessive heat, excessive cold, and operator error. Ways to detect a gas leak include cloth streamers, leak detectors, scent indicators, hearing a leak, or observing localized corrosion.

77. A: The main types of mechanisms usually responsible for causing structural failure include extreme temperatures, corrosion, tension, buckling, shearing, compression-bearing fatigue, instability, and material creep. Stress is not a mechanism responsible for causing failure as it is simply the amount of force per unit area. Stress is often confused with strain, the amount of deformation a material experiences in response to a stress.

78. B: High-pressure fluids are used in such devices as fire hoses, fuel-injection mechanisms, and paint sprayers. Associated hazards usually include injection injuries, pressurized gas-impact injuries, and line whipping.

79. C: Installed ramps must have a slope of less than 15 degrees for the general public and less than 11 degrees for handicapped access.

80. C: Freeze plugs operate under the principal of water or water-based liquids expanding as they approach their freezing point temperatures. As a liquid gets colder and expands, the freeze plug ultimately allows the liquid to drain, as necessary, from its container.

81. A: Typically employed controls for preventing materials-handling accidents include creating <u>wide</u> (not narrow) passageways and lanes for movement of materials, training workers to properly use hand signals, training workers on proper heavy-lifting techniques, implementing traffic controls to keep lift areas clear of people, and regularly maintaining and inspecting handling equipment.

82. B: The "angle of repose" is defined as the natural angle that soil (or other earthen material) forms when it is piled up or when it collapses.

83. B: The "range" of a data set is the numerical difference between the highest and lowest values in the set. Thus, 44 – 3 = 41.

84. C: A "leading indicator" is conventionally defined as an objective measure that is used to assess proactive actions taken to improve organizational performance.

85. D: Workers' compensation laws have several primary objectives. These include saving workers the nuisances of litigation, preventing injured workers from having to solicit charities, empowering employers to develop procedures that prevent and reduce accidents, replacing lost worker income, providing prompt medical treatment for sustained injuries, providing workers with rehabilitation options so that they can return to work more quickly, and encouraging accident investigations to ultimately prevent similar events from occurring again. The goal of finding fault (culpability) for accident situations is not a primary objective of workers' compensation laws.

86. A: When a near-miss occurs at a work site, the most appropriate response is to treat it as if it were an actual injury event, by completing a root-cause analysis and developing action plans to prevent potential recurrences.

87. B: Routinely scheduled inspections are an integral part of an effective occupational health and safety system. As such, the frequency of inspections must be aligned with the degree of risk posed by an operation, in conjunction with regulatory requirements. Inspections that are not performed at adequate frequencies are likely indicative of a substandard occupational health and safety program.

88. C: Noneconomic costs associated with occupational injuries can have serious impacts on a subject organization. Such costs may include worker morale, absenteeism, and poor productivity.

89. B: Several safeguards can be employed to protect workers from accidents involving machines or tools. These include ensuring a regular inspection and maintenance schedule is executed, ensuring that workers have no loose clothing, jewelry, or hair while using machines and tools in which these items may become caught, ensuring the proper functioning of machine and tool guards, and ensuring proper training for all who may use the subject machines or tools. Ensuring proper lockout and tagout should only be done when machines and tools are being set up, cleaned, or maintained so as to prevent an inadvertent turn-on or energization event. Lockout and tagout should never be employed while a machine is in use.

90. D: The "variance" of a statistical population is defined as the square of the standard deviation of that population.

91. D: Environmental hazards can include such phenomena as pressure, heat, chemicals, noise, light, and radiation. Such hazards can often have latent characteristics associated with them, in that their effects are often not observed directly, may only appear over time, or may not be apparent at all until years after exposure. Such potential effects may include illnesses, hearing loss, burns, and irritability.

92. A: There are three fundamental modes of air-supply equipment used in the workplace for ensuring good-quality breathable air. These are self-contained respirators (such as SCBAs), supplied-air respirators (such as air-line modules, hose masks, air-supplied suits, and hoods), and air-purifying respirators (such as a full-face or half-face respirator).

93. B: The primary factors to keep in mind when evaluating potential chemical hazards are: (1) compounds or materials that aren't normally hazardous on their own may become hazardous when combined with other substances or when used in certain ways; and (2) compounds or materials that are known to be generally hazardous may not be "dangerous" below certain threshold concentrations.

94. D: Personal protective equipment (PPE) serves as a barrier between a worker and a hazard, but it does not remove the hazard. PPE should never be considered a means for removing hazards or for viewing a hazardous environment as a "safe" environment.

95. C: Hazardous materials can enter the body through the following pathways: inhalation, ingestion, absorption through the skin or mucus membranes, or directly through an open cut or wound.

96. A: A fire caused by flammable liquids may be effectively extinguished via the use of halon gas. Halon is also effective on electrical fires. Water should <u>not</u> be used to treat a flammable-liquid fire or electrical fire. Spreading of flames and electrical hazards may result from the attempted use of water in these instances, respectively.

97. B: Worksite sanitation essentially deals with ensuring a worksite or facility is clean and free of germs. It includes ensuring that food areas are clean enough for healthy preparation and consumption, ensuring a safe drinking-water-supply system, ensuring clean and working toilets and sinks, and providing control for insects and rodents.

98. C: A repetitive strain injury usually results from long-term, cumulative trauma or stress to ligaments, tendons, muscles, nerves, and joints. Hands, arms, shoulders, and necks are usually the most susceptible areas to such injuries. Typical conditions associated with repetitive strain injuries include tendonitis, carpal tunnel syndrome, trigger finger, and bursitis.

99. C: Manlifts are inherently characterized by a number of potential hazards if proper care is not exercised during their use. Such primary hazards may include people potentially falling off the manlift platform, people becoming caught between the edge of a manlift and floor openings through which it passes, and operation by untrained personnel.

100. B: When developing a maintenance checklist, manufacturer's specifications and product literature should be the primary references from which checklist items are developed, including recommended maintenance frequencies.

101. C: There are several measures that may be employed in the workplace for controlling exposures to chemical hazards. These include PPE, work modifications, personal hygiene, and reducing worker exposure time.

102. D: Relief valves are overpressure devices that open when upstream pressures are higher than pre-set values. They are typically only used for liquids; the extent to which they open up depends on the amount of system overpressure, and they close automatically when system pressure returns to normal.

103. C: There a number of hazards that inherently exist with the use of hand vehicles such as dollies, carts, wheelbarrows, and hand trucks. These may include running into walls, equipment, and people, poor visibility when loads are stacked too high, jamming of hands against walls, and the tipping over of loads due to instability.

104. A: A "lagging indicator" is conventionally defined as a metric or occurrence that has transpired after a workplace injury has occurred.

105. B: Safeguarding practices for hand tools can be implemented in a variety of forms. For example, bent handles can help prevent tissue-compression and repetitive-motion disorders. Other beneficial practices also include using axes and hatchets with longer handles, and using high-friction plastic handles to facilitate a tighter grip.

106. B: Reducing potential hazards associated with ladder use can entail a number of actions, including ensuring that a ladder's rungs are slip-resistant, ensuring that no electrical conductors are nearby (especially when using a metal ladder), inspecting ladders for damage, and anchoring a ladder to a robust support structure for stability. Ladders should also be placed far enough from a wall so that the <u>arch of the foot</u> (not just the toes) can fit comfortably on the rungs.

107. B: Hearing-protective PPE primarily consists of earplugs and earmuffs (muffs). Muffs are usually more effective at blocking out noise than earplugs; however, both items may be worn in tandem for extremely noisy environments. In the latter case, it is important to have a robust hand-signal system in place to compensate for the potential absence of clear auditory communication.

108. D: Microwaves <u>are not</u> a form of ionizing radiation; they excite electrons but do not strip or dislocate them from their shells (hence, the process of "ionization"). Forms of ionizing radiation include alpha particles, beta particles, gamma rays, X-rays, and neutrons. Ionizing radiation is the most dangerous type of radiation because it can chemically change particles, which can result in mutations and cancer.

109. D: Reducing potential hazards associated with scaffold use can entail a number of actions, such as ensuring that the subject scaffold is appropriately rated for the weights of loads it is to handle, regularly ensuring that outrigger beams are not damaged, and regularly ensuring that vital hardware such as planks, bolts, ropes, bracing, and clamps are not damaged. Lanyards and safety belts should be used on scaffolds more than 10 feet above the ground.

110. A: Tripping hazards around a worksite can be significantly reduced in a number of ways. Examples include avoiding one- and two-step elevation changes, ensuring that flooring-transition areas are level, taping down electrical cords if they cross walking paths, following good housekeeping practices, and posting warnings where there are floor elevation changes and potential dangers.

111. B: Soil and other earthen materials contain levels of radium and radon, as these are naturally occurring substances in the ground. Both radium and radon emit ionizing radiation in the form of alpha particles upon their decay. Radium and radon concentrations can vary depending on geographic region, weather, hydrology, and degree of ventilation. Radiofrequency (RF) emitters are not considered a source of ionizing radiation.

112. D: To effectively follow up on and correct audit findings, subject items should be grouped, as appropriate, into the following categories: (1) nonconformances, (2) areas of concern, and (3) opportunities for improvement.

113. C: The purpose of tracking incidents and collecting data is to use that data as a continuous improvement tool that provides feedback on performance and resultantly guides active response.

114. C: A well-stocked first-aid kit should include the following items: disposable sterile gloves, mouthpieces, scissors, tweezers, rubbing alcohol, sterile gauze dressings, at least two sterile eye dressings, triangular bandages, rolled bandages, safety pins, cleansing wipes, sticky tape, antiseptic cream, antihistamine tablets, distilled water, and an eye wash or bath.

115. A: The U.S. Department of Transportation divides explosives into three (3) separate categories. Class A is the most dangerous due to its detonating hazard, and may include substances such as black powder, nitroglycerin, TNT, and certain ammunition. Class B is less dangerous than Class A but is still hazardous due to functioning by rapid combustion instead of detonation; signal devices, smokeless powders, and fireworks are examples of this class. Class C is the least dangerous due to its containing very restricted amounts of Class A or B materials; examples of this class include certain types of fireworks and manufactured items.

116. C: According to the "Human Factors Theory of Accident Causation," accidents are primarily caused by human error resulting from the following categories: (1) overload, (2) inappropriate response, and (3) inappropriate activities. Overload can occur when workers' responsibilities are greater than their capacities; inappropriate response refers to any actions (before or after an accident) that cause the accident itself or make it more severe; and inappropriate activities point to jobs or tasks for which a worker is not adequately trained.

117. B: Primary functions of machine guards are to keep body parts, clothing, or hair from coming into contact with hazardous machine parts, and to prevent flying debris from striking people. Secondary functions include muffling machine noise, capturing dust, and containing or exhausting contaminants.

118. C: Safety programs that reduce the frequency and severity of accidents will likewise reduce workers' compensation claims.

119. D: The science of probabilistic risk assessment defines "risk" as the product of probability and consequence. For example, if an accident has an annual probability of occurrence equal to 0.00001/year and the associated consequence is five (5) expected fatalities resulting from that occurrence, then the resulting annualized risk for such an occurrence is expected to be 0.00005 fatalities/year.

120. B: For a standard (normal) distribution of data, 68 percent of values will fall within 1 standard deviation of the mean, 95 percent of values will fall within 2 standard deviations of the mean, and 99.7 percent of values will fall within 3 standard deviations of the mean. Such a distribution is typically known as a "bell curve" because its graphical shape resembles that of a bell.

121. C: Basic "book value" at a point in time is calculated as initial purchase price minus accumulated depreciation at that point in time.

122. A: The four crucial elements for implementing a successful personal-protective-equipment (PPE) program are: (1) selection, (2) management, (3) use, and (4) maintenance. Permitting workers to participate in equipment selection can help them "buy into" a PPE program, which increases the likelihood that they will consistently and properly use their equipment when it is needed.

123. C: "Latency period" is defined as the amount of time between a hazardous exposure and the observable health effects due to that exposure. Latency periods can be quite brief, as with a chemical burn, or can be significantly delayed, such as with the onset of cancer from a high dose of radiation.

124. A: An acute exposure refers to only a single exposure (usually high in magnitude) to a hazard, whereas chronic exposure refers to repeated contact or interaction with a hazard. Significant health effects may ultimately manifest from either type of exposure.

125. B: An "asphyxiant" is defined as a material that displaces oxygen in air, thus interfering with normal breathing and sufficient oxygen transport in the bloodstream.

126. D: Hazardous exposures in high-enough quantities can result in "local effects" or "system effects" to the human body. Local effects typically affect or injure the eyes, skin, or respiratory system, whereas systemic effects usually affect or damage organs and body functions.

127. D: There are three degrees of burns to which the human body is susceptible: first, second, and third. A first-degree burn is superficial, with some reddening and pain, with healing usually taking around a week's time; a second-degree burn is deep, with blisters and pain, with healing usually taking around a few weeks' time; and a third-degree burn is characterized by all skin layers virtually being destroyed, with healing potentially taking several months. Due to the potential destruction of nerve endings in the skin from a third-degree burn, it is often less painful than a second-degree burn.

128. B: The most common type of workplace injury are sprains and strains, followed by bruises, cuts, and fractures. Most subject injuries occur due to overexertion, slips, trips, falls, and impacts.

129. A: The lower back is the most frequently injured part of the body (mainly due to poor lifting techniques) in the workplace. After the back, the knees, ankles, fingers, and arms are the most commonly injured.

130. D: An "overexposure" to hazardous materials (i.e., chemicals or radiation) is not overly common in the workplace, primarily due to the use of engineering controls, administrative controls, personal protective equipment, safety program implementation, and adherence to OSHA requirements.

131. A: Workplace incidents and accidents are usually attributable to one or more of the following causal factors: (1) mismatch or overload, (2) unsafe conditions, (3) unsafe acts, (4) systems failure, (5) traps, (6) personal beliefs and feelings, and (7) a decision to work unsafely. Morale, in and of itself, is not usually a prime causal factor that directly leads to an event.

132. C: There are three types of engineering controls that can be implemented in the workplace for protecting people from hazardous exposures: (1) substitution – opting to replace a hazardous material with a less-hazardous or non-hazardous material; (2) isolation – creating a physical barrier between workers and hazards; and (3) ventilation – continually replacing hazardous or contaminated air with fresh air.

133. B: A lanyard's primary function within a fall-protection system is to connect a safety harness to an anchoring point. Lanyards absorb energy, so they reduce the impact load on a person when a fall is arrested.

134. A: Rigging is the connecting apparatus affixed between a lifting device and a load. Such examples include ropes, chains, cables, and slings.

135. A: Personal protective equipment (PPE) provides protection from hazardous materials, heat, sparks, fire, biohazards, noise, impact injuries, flying debris, and poor visibility. Examples of PPE include hardhats, earplugs, earmuffs, eye goggles, face shields, coats, aprons, coveralls, hoods, booties, gloves, steel-toed shoes, respirator masks, full-body suits, fire suits, rainwear, high-visibility clothing and vests, and certain flotation devices. Standard sunglasses with UV coating are not considered PPE, though there are certain goggles, shields, and masks that are designed with such coatings to protect workers' eyes from potential UV exposures in the workplace (such as from welding-torch light).

136. B: On average, statistics show that as many as 8 percent of workplace incidents involving lost time are related to hand-tool injuries.

137. D: A winch is a device that is used for hauling or lifting, and consists of a rope, cable, or chain winding around a horizontal rotating drum.

138. C: There are three separate classifications for flammable liquids: IA, IB, and IC. Methyl ethyl ketone is a Class "IB" flammable liquid. Class IB flammable liquids are defined as those that have flash points below 73°F (22.8°C) and boiling points at or above 100°F (37.8°C). In contrast, Class IA liquids have flash points below 73°F (22.8°C) and boiling points below 100°F (37.8°C), and, Class IC liquids have flash points at or above 73°F (22.8°C) and boiling points below 100°F (37.8°C).

139. D: For working in areas with high concentrations of particulate hazards (e.g., lead, beryllium), full-face respirators are an appropriate choice for respiratory PPE. In addition to their high-efficiency filtration, they also provide coverage to the eyes.

140. B: Per OSHA regulations, a respirator should be equipped with a green-colored canister for providing protection from ammonia gas.

141. A: Construction documentation pertaining to project design requirements typically includes such items as plans, specifications, shop drawings, shop-drawing logs, request-for-information logs, submittal logs, and change-order logs. In contrast, purchase orders are conventionally associated with the front end of contract proceedings and are primarily fiscal-based documents.

142. C: Typical means of enhancing construction ergonomics may include such equipment as adjustable scaffolding, reduced vibration power tools, and motorized screeding equipment.

143. B: It is a best practice for construction contractors to maintain a daily diary that keeps accurate track of and accountability for on-site employees, subcontractors, visitors, deliveries, weather, project discrepancies, and important meetings and conversations that are held.

144. C: The American National Standards Institute (ANSI) and the National Commission for Certifying Agencies (NCCA) are the two primary U.S.-national accrediting agencies responsible for overseeing organizations that offer crane operator certification programs.

145. B: Demographic data or statistics should not be used as a metric for determining whether worker competencies are commensurate with responsibilities. Worker competency levels as they relate to qualifications and responsibilities should be primarily based on work experience, education and training, and communication ability.

146. A: OSHA 29 CFR 1926.21 specifically addresses the domain of "Construction Worker Safety Training and Education." Part 1926, in its entirety, covers the entire spectrum of construction-related OSHA regulations.

147. D: OSHA 29 CFR 1926.1103 (re: 29 CFR 1910.1003) covers the thirteen (13) primary heavily carcinogenic substances that may be found in the construction workplace. These include:

4-nitrobiphenyl, alpha-naphthylamine, methyl chloromethyl ether, 3, 3'-dichlorobenzidine, bis-chloromethyl ether, beta-naphthylamine, benzidine, 4-aminodiphenyl, ethyleneimine, beta-propiolactone, 2-acetylaminofluorene, 4-dimethylaminoazo-benzene, and N-nitrosodimethylamine. Ammonium hydroxide (a.k.a., ammonia-water) is not categorized as a carcinogen, and though toxic, is typically utilized as a main active ingredient in cleaning solutions.

148. B: A certified industrial hygienist (CIH) would likely be the most qualified individual to measure a "variety of hazards" in the workplace and resultantly provide recommendations on how to effectively mitigate such hazards. A certified health physicist (CHP) is primarily a subject-matter expert regarding radiological hazards and doses; a certified associate safety professional's (ASP's) breadth of expertise and rigor of training and certification are typically less than a CIH's; and an industrial (professional) engineer's main area of focus is usually innovating processes and systems that improve quality and productivity.

149. D: Rubber-tired self-propelled scrapers, rubber-tired front-end loaders, rubber-tired dozers, wheel-type agricultural and industrial tractors, crawler tractors, crawler-type loaders, and motor graders that are used in construction work shall all be equipped with rollover protective structures.

150. C: The BCSP requirements for CHST recertification are the reporting of activities every five years, and a minimum of 20 recertification points. Points may be accumulated through continuing education, training, work practices, memberships in safety organizations, publications, presentations, and patents.

151. B: Per OSHA 29 CFR 1926.56, general construction worksite (plant) areas and shops shall be illuminated at a light intensity of no less than 10 foot-candles.

152. C: The following three (3) resources are specifically called out for reference within the realm of chemical hazard communication protocol, per OSHA 1910.1200: the United Nations' Globally Harmonized System of Classification and Labelling of Chemicals (GHS), (Material) Safety Data Sheets, and guidance per the U.S. EPA's Toxic Substances Control Act. CDC's National Institute for Occupational Safety and Health (NIOSH) independently publishes a "recommended exposure limit" (REL) compendium for chemicals.

153. A: Per OSHA 29 CFR 1926.96, safety-toe footwear for construction employees shall meet the requirements and specifications stated within the ANSI Standard for Men's Safety-Toe Footwear, Z41.1-1967.

154. D: Protective-certified headwear (hardhats, helmets) must be worn in any construction area that entails hazards from direct impacts, falling objects, flying objects, burns, and electric shock. Such headwear must be robustly constructed of materials to sufficiently absorb impact energies, be flame-resistant, and provide electrical insulation.

155. C: Per OSHA 29 CFR 1926.106, construction employees working over or near water, where the danger of drowning exists, shall be provided with U.S. Coast Guard-approved life jackets or buoyant work vests.

156. B: A "Type-A" rated fire extinguisher is effective against a trash-wood-paper-based fire, and is symbolically denoted by a green triangle. A "Type-B" extinguisher is effective against flammable or combustible liquids such as gasoline, kerosene, grease, and oil, and is denoted by a red square. A "Type-C" **extinguisher is effective against** fires involving electrical equipment, such as appliances, wiring, circuit breakers, and outlets, and is denoted by a blue circle.

157. C: Per OSHA 29 CFR 1926.404, all 120-volt single-phase 15- and 20-ampere receptacle outlets on construction sites, which are not a part of the permanent wiring of a building or structure and which are in use by employees, shall have approved ground-fault circuit interrupters for personnel protection.

158. A: Per OSHA 29 CFR 1926.441, facilities for quick drenching of the eyes and body shall be provided within 25 feet (not 100 feet) of battery handling areas. In addition, unsealed batteries shall be located in enclosures with outside vents or in well-ventilated rooms, and shall be arranged so as to prevent the escape of fumes, gases, or electrolyte spray. Moreover, when batteries are being charged, vent caps must remain in place.

159. C: Construction worksite fall-protection training must cover, as applicable, the proper use and operation of guardrail systems and controlled access zones, as well as personal fall-arrest systems, safety net systems, warning line systems, safety monitoring systems, or any other protection mechanisms that are to be employed at the worksite.

160. B: Conveyor systems shall be equipped with an audible warning signal to be sounded immediately before starting up the conveyor. In addition, emergency-stop switches shall be arranged so that the conveyor cannot be started again until the actuating stop switch has been reset to the "on" position, and all conveyors shall meet applicable requirements for design, construction, inspection, testing, maintenance, and operation, as prescribed in the ANSI B20.1-1957. If a conveyor operator station is at a remote point, a stop-switch for the conveyor motor <u>shall be</u> provided at the motor location.

161. A: Employees engaged in site-clearing activities shall be protected from the hazards of irritant and toxic plants and flora. As such, they are to be suitably instructed in first-aid treatment, and all rider-operated equipment shall be furnished with rollover guards as well as overhead- and rear-canopy guards.

162. D: The three fundamental modes of "shoring" used in trenching and other excavating work are aluminum hydraulic shoring, timber shoring, and pneumatic shoring.

163. D: The following construction tools and machines are generally considered to be "high(est)-risk" for potentially causing severe injuries: circular saws, table saws, nail guns, power drills, chainsaws, air compressors, wood chippers, and ladders. Although conveyors also maintain inherent hazards, they are generally not considered "high risk."

164. D: The following types of concrete or concrete-related operations are most commonly used in the construction industry: shotcrete, precast concrete, and fiber-reinforced concrete.

165. B: During demolition preparation activities, if any hazardous substance(s) is/are apparent or suspected, testing and purging shall be conducted, with the hazard being subsequently eliminated before actual demolition activities are commenced.

166. A: Per OSHA 1926.961, the following are considered typical standard operating procedures for the re-energization of lines and equipment: ensuring that all crews working on lines or equipment release their clearances, ensuring that all employees are physically clear of lines and equipment, and ensuring that all protective tags are removed from the point of disconnection.

167. B: OSHA regulations maintain several safety requirements for temporary stairways installed for construction activities These include that stairways shall have landings of not less than 30 inches (76 cm) in the direction of travel and must extend at least 22 inches (56 cm) in width at every 12 feet (3.7 m) or less of vertical rise; that variations in riser height or tread depth shall not be over 1/4 inch (0.6 cm) in any stairway system; that where doors or gates open directly on a stairway, a platform shall be provided, and the swing of the door shall not reduce the effective width of the platform to less than 20 inches (51 cm); and finally, that stairs must be installed between 30 and 50 degrees (not 20 and 55 deg.) from horizontal.

168. B: The term "fit-testing" is used when referring to respirator PPE. A "fit-test" tests the seal between the respirator's face-piece and the user's face. It normally takes about 15-20 minutes to complete and is performed at least annually. After passing a fit-test with a respirator, a worker must use the exact same make, model, style, and size respirator on the job. A fit-test should not be confused with a user seal-check, which ensures that the respirator is properly seated to the face or needs to be readjusted each time the equipment is donned.

169. D: Whenever there is a concern as to safety, crane and derrick operators must have the authority to stop and refuse to handle loads until a qualified person has determined that safety has been resoundingly assured.

170. C: A masonry-project "limited-access zone" is required to be equal to the height of the wall to be (re)constructed plus an additional four feet, and shall run the entire length of the wall.

171. D: A "surcharge load" is defined as a material or physical mechanism that can increase the likelihood of a trench cave-in or collapse. Examples of surcharge loads are heavy equipment situated or stored close to trench walls, spoil-piles amassed close to trench walls, and personnel walking near the trench. Workmen inside the trench would not increase the likelihood of collapse.

172. B: Low-compressive-strength materials are those that can easily come apart or separate due to external forces or even just from the force of gravity alone. Examples of low-compressive-strength materials include sand, gravel, and unstable rock. In contrast, examples of high-compressive-strength materials include clay, clay loam, and cemented soils that are highly resistant to crumbling.

173. D: Electric shock can potentially result in a variety of injuries, including cardiac arrest due to electrical effects on the heart; muscle, nerve, and tissue destruction from current passing through the body; and, most commonly, thermal burns.

174. C: There are several measures or techniques that are regularly employed against controlling static electricity in the workplace; these include humidification, incorporation of additives to reduce charge accumulation, electrical bonding, and electrical grounding.

175. A: A steel-erection contractor shall not erect steel unless it has received written notification that the concrete in the footings, piers, and walls, or the mortar in the masonry piers and walls, has attained, on the basis of an appropriate ASTM standard test method of field-cured samples, either 75 percent of the intended minimum compressive design strength or sufficient strength to support the loads imposed during steel erection.

176. B: To ensure scaffold safety, the following minimum requirements are deemed necessary for an employee to be on a moving scaffold: outriggers are installed on both sides of the scaffold; the scaffold height-to-base ratio is 2:1; workers are only located at scaffold positions inside the wheels; scaffold platforms overlap by at least 12 inches over their supports, unless restrained otherwise; and a competent (qualified) person is on site to supervise scaffold movements.

177. C: The only time a body belt may be used where there may be a fall hazard is when a construction worker is using a positioning device.

178. C: A "warning line system" is conventionally defined as an erected barrier on a rooftop to warn workers they are approaching an unprotected side or edge.

179. A: "Orthostatic incompetence" (a.k.a., suspension trauma) is defined as the experienced medical effects of being immobilized or suspended in a vertical position. Such effects may include dizziness, sweatiness, increased pulse and respiration, sudden drop in blood pressure, unconsciousness, and possible death.

180. C: Welding, cutting, and brazing operations can entail many potential hazards, which may include severe burns, air contaminants, potential eye damage from UV light, explosions, and repetitive-strain injuries.

181. B: Database software tools would be most appropriate for safety-program recordkeeping and audit tracking. Spreadsheet software programs could also be used for such purposes, but likely to more limited capacities than their database counterparts.

182. B: Security and privacy of worker personal health records is covered and protected by law under the Health Insurance Portability and Accountability Act (HIPAA).

183. D: Acetylene gas is commonly used in oxygen-fuel welding and cutting. It is the most common gas used for fueling cutting torches in the construction industry. When mixed with pure oxygen in a cutting-torch assembly, an acetylene flame can potentially reach over 5,700°F. It is very sensitive to conditions such as excess pressure, excess temperature, static electricity, and mechanical shock. Because of its unstable nature, it must be stored under special conditions. This is often accomplished by dissolving the acetylene in liquid acetone and then storing the mixture in a cylinder filled with porous cementitious material.

184. C: Hazardous occurrences often result in the workplace from the misuse or poor maintenance of certain hand tools. Such misuse can include using these tools in manners they were not fashioned or designed for, including taking shortcuts. Such examples include eye injuries caused by inappropriately using a screwdriver as a chisel; pieces of wooden handles on hammers or axes breaking off and striking an individual; and the inadvertent springing open of a wrench's jaws while in (over-)use. The stripping or failure of a pliers' set or connecting screw is not typically associated with misuse but rather more indicative of a manufacturer's defect.

185. A: "White finger" syndrome (a.k.a., Raynaud's syndrome, hand-arm vibration syndrome, or "dead finger") in the construction workplace is typically caused by the continuous use of vibrating hand-held machinery. It is a disorder that affects the nerves, blood vessels, muscles, and joints of the hand, wrist, and arm, and poses the greatest risk of occurrence at vibration frequencies between 50 and 150 Hz.

186. B: The MOST vital safety rule to always remember when using a portable circular saw is to NEVER hold or balance a piece of material in one hand while trying to operate the saw with the other. In addition, other important safety considerations include always wearing adequate eye and face protection, always using a sharp-enough blade, and always ensuring that the saw's switch actuates properly.

187. D: "Lifting frequency" for construction workers should be tallied as the average number of lifts a worker performs over a 15-minute period.

188. C: A number of compressed-air (pneumatic) tools are regularly used on a construction site. Nail guns, chippers, sanders, saws, drills, paving breakers, diggers, jackhammers, rivet busters, chipping hammers, and tampers are most regularly utilized. Pneumatic tools often entail additional hazards that their non-pneumatic counterparts do not. Such examples include additional forces or inertia due to high air pressures, as well as noise levels, pollutant emissions, shock potential, whipping-hose dangers, need for additional eye protection, and chilled-air emissions.

189. B: Whenever an OSHA inspection occurs, the following items are deemed mandatory: (1) the employer must be advised by the compliance officer as to the reason for the inspection; (2) the employer must accompany the compliance officer on the inspection; (3) the compliance officer must show official identification; and (4) the employer must be assured by the compliance officer that any trade secrets observed during the inspection will remain confidential.

190. A: A severity rate (SR) of 52.29 is calculated for the scenario of three recordable injuries with one of these resulting in 65 days of lost time, and a total number of workforce hours worked equal to 248,620.

SR = (number of lost workdays × 200,000) ÷ (total number of hours worked)

$$= (65 \times 200{,}000) \div (248{,}620)$$
$$= 52.29$$

191. B: The trigonometric "tangent" of angle ∠ABC is equal to "b/a" ("opposite segment divided by adjacent segment" or "height divided by base").

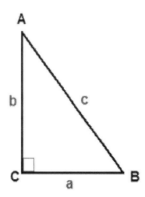

192. C: A strong acid typically has a pH around 1, and a strong base typically has a pH around 14. Strong acids and bases are very caustic and usually present numerous potential hazards to human health if exposure occurs via ingestion, inhalation, or externally. In addition to their causticity, most act as chemical poisons as well. Examples of strong acids include hydrochloric acid, sulfuric acid, and nitric acid. Examples of strong bases include sodium hydroxide, potassium hydroxide, and lithium hydroxide.

193. C: There are numerous agents or substances in the workplace to which workers may potentially be exposed that are tied to the manifestation of specific diseases or disorders. Such agents and substances include beryllium, which can lead to berylliosis; silica, which can lead to silicosis; and cadmium, which can lead to a number of serious systemic conditions. Argon, however, is an inert gas.

194. B: The primary types of ventilation-measurement equipment used in the workplace are pitot tubes, rotating vane anemometers, and thermal anemometers.

195. C: There are several potential violation categories that exist under OSHA. These include a willful violation (a violation that an employer intentionally and knowingly commits), a serious violation (a violation that entails a substantial possibility that death or serious physical harm can result), a failure-to-abate violation (a failure to correct a prior violation), a repeated violation (a violation whereby upon [re-]inspection is found to be very similar to a previous violation), and an other-than-serious violation (a violation that has a direct relationship to job safety and health, but in all likelihood would not cause death or serious injury).

196. A: Per OSHA's audiometric testing standards, a baseline audiogram must be conducted within six months of a first noise exposure greater than 85 dBA.

197. C: There are several cold-stress-related conditions, injuries, or illnesses that may arise in the workplace, including trench foot, chilblains, frostbite, and <u>hypo</u>thermia. <u>Hyper</u>thermia, contrastingly, is caused by heat stress.

198. D: The term "TLV," in regard to hazardous chemical exposure, stands for "threshold limit value." It is defined as a consensus level to which a worker can be exposed day after day for a working lifetime (to a subject chemical) without the likely manifestation of adverse health effects. "TLV" is a reserved term of the American Conference of Governmental Industrial Hygienists (ACGIH).

199. D: A multimeter is a device that can be used to measure electrical voltage, current, or resistance. It can be a hand-held device useful for basic fault-finding and field-service work, or a bench instrument that can measure to a very high degree of accuracy. It can be used to troubleshoot electrical problems in a wide array of industrial systems such as electronic equipment, motor controls, power supplies, and wiring systems.

200. A: There are typically several options available that may be employed for the delivery of safety training in the workplace; primary of these are instructor-led training, self-paced training (including online), and structured on-the-job training.

How to Overcome Test Anxiety

Just the thought of taking a test is enough to make most people a little nervous. A test is an important event that can have a long-term impact on your future, so it's important to take it seriously and it's natural to feel anxious about performing well. But just because anxiety is normal, that doesn't mean that it's helpful in test taking, or that you should simply accept it as part of your life. Anxiety can have a variety of effects. These effects can be mild, like making you feel slightly nervous, or severe, like blocking your ability to focus or remember even a simple detail.

If you experience test anxiety—whether severe or mild—it's important to know how to beat it. To discover this, first you need to understand what causes test anxiety.

Causes of Test Anxiety

While we often think of anxiety as an uncontrollable emotional state, it can actually be caused by simple, practical things. One of the most common causes of test anxiety is that a person does not feel adequately prepared for their test. This feeling can be the result of many different issues such as poor study habits or lack of organization, but the most common culprit is time management. Starting to study too late, failing to organize your study time to cover all of the material, or being distracted while you study will mean that you're not well prepared for the test. This may lead to cramming the night before, which will cause you to be physically and mentally exhausted for the test. Poor time management also contributes to feelings of stress, fear, and hopelessness as you realize you are not well prepared but don't know what to do about it.

Other times, test anxiety is not related to your preparation for the test but comes from unresolved fear. This may be a past failure on a test, or poor performance on tests in general. It may come from comparing yourself to others who seem to be performing better or from the stress of living up to expectations. Anxiety may be driven by fears of the future—how failure on this test would affect your educational and career goals. These fears are often completely irrational, but they can still negatively impact your test performance.

Review Video: 3 Reasons You Have Test Anxiety
Visit mometrix.com/academy and enter code: 428468

Elements of Test Anxiety

As mentioned earlier, test anxiety is considered to be an emotional state, but it has physical and mental components as well. Sometimes you may not even realize that you are suffering from test anxiety until you notice the physical symptoms. These can include trembling hands, rapid heartbeat, sweating, nausea, and tense muscles. Extreme anxiety may lead to fainting or vomiting. Obviously, any of these symptoms can have a negative impact on testing. It is important to recognize them as soon as they begin to occur so that you can address the problem before it damages your performance.

> **Review Video: 3 Ways to Tell You Have Test Anxiety**
> Visit mometrix.com/academy and enter code: 927847

The mental components of test anxiety include trouble focusing and inability to remember learned information. During a test, your mind is on high alert, which can help you recall information and stay focused for an extended period of time. However, anxiety interferes with your mind's natural processes, causing you to blank out, even on the questions you know well. The strain of testing during anxiety makes it difficult to stay focused, especially on a test that may take several hours. Extreme anxiety can take a huge mental toll, making it difficult not only to recall test information but even to understand the test questions or pull your thoughts together.

> **Review Video: How Test Anxiety Affects Memory**
> Visit mometrix.com/academy and enter code: 609003

Effects of Test Anxiety

Test anxiety is like a disease—if left untreated, it will get progressively worse. Anxiety leads to poor performance, and this reinforces the feelings of fear and failure, which in turn lead to poor performances on subsequent tests. It can grow from a mild nervousness to a crippling condition. If allowed to progress, test anxiety can have a big impact on your schooling, and consequently on your future.

Test anxiety can spread to other parts of your life. Anxiety on tests can become anxiety in any stressful situation, and blanking on a test can turn into panicking in a job situation. But fortunately, you don't have to let anxiety rule your testing and determine your grades. There are a number of relatively simple steps you can take to move past anxiety and function normally on a test and in the rest of life.

> **Review Video: How Test Anxiety Impacts Your Grades**
> Visit mometrix.com/academy and enter code: 939819

Physical Steps for Beating Test Anxiety

While test anxiety is a serious problem, the good news is that it can be overcome. It doesn't have to control your ability to think and remember information. While it may take time, you can begin taking steps today to beat anxiety.

Just as your first hint that you may be struggling with anxiety comes from the physical symptoms, the first step to treating it is also physical. Rest is crucial for having a clear, strong mind. If you are tired, it is much easier to give in to anxiety. But if you establish good sleep habits, your body and mind will be ready to perform optimally, without the strain of exhaustion. Additionally, sleeping well helps you to retain information better, so you're more likely to recall the answers when you see the test questions.

Getting good sleep means more than going to bed on time. It's important to allow your brain time to relax. Take study breaks from time to time so it doesn't get overworked, and don't study right before bed. Take time to rest your mind before trying to rest your body, or you may find it difficult to fall asleep.

> **Review Video: The Importance of Sleep for Your Brain**
> Visit mometrix.com/academy and enter code: 319338

Along with sleep, other aspects of physical health are important in preparing for a test. Good nutrition is vital for good brain function. Sugary foods and drinks may give a burst of energy but this burst is followed by a crash, both physically and emotionally. Instead, fuel your body with protein and vitamin-rich foods.

Also, drink plenty of water. Dehydration can lead to headaches and exhaustion, especially if your brain is already under stress from the rigors of the test. Particularly if your test is a long one, drink water during the breaks. And if possible, take an energy-boosting snack to eat between sections.

> **Review Video: How Diet Can Affect your Mood**
> Visit mometrix.com/academy and enter code: 624317

Along with sleep and diet, a third important part of physical health is exercise. Maintaining a steady workout schedule is helpful, but even taking 5-minute study breaks to walk can help get your blood pumping faster and clear your head. Exercise also releases endorphins, which contribute to a positive feeling and can help combat test anxiety.

When you nurture your physical health, you are also contributing to your mental health. If your body is healthy, your mind is much more likely to be healthy as well. So take time to rest, nourish your body with healthy food and water, and get moving as much as possible. Taking these physical steps will make you stronger and more able to take the mental steps necessary to overcome test anxiety.

> **Review Video: How to Stay Healthy and Prevent Test Anxiety**
> Visit mometrix.com/academy and enter code: 877894

Mental Steps for Beating Test Anxiety

Working on the mental side of test anxiety can be more challenging, but as with the physical side, there are clear steps you can take to overcome it. As mentioned earlier, test anxiety often stems from lack of preparation, so the obvious solution is to prepare for the test. Effective studying may be the most important weapon you have for beating test anxiety, but you can and should employ several other mental tools to combat fear.

First, boost your confidence by reminding yourself of past success—tests or projects that you aced. If you're putting as much effort into preparing for this test as you did for those, there's no reason you should expect to fail here. Work hard to prepare; then trust your preparation.

Second, surround yourself with encouraging people. It can be helpful to find a study group, but be sure that the people you're around will encourage a positive attitude. If you spend time with others who are anxious or cynical, this will only contribute to your own anxiety. Look for others who are motivated to study hard from a desire to succeed, not from a fear of failure.

Third, reward yourself. A test is physically and mentally tiring, even without anxiety, and it can be helpful to have something to look forward to. Plan an activity following the test, regardless of the outcome, such as going to a movie or getting ice cream.

When you are taking the test, if you find yourself beginning to feel anxious, remind yourself that you know the material. Visualize successfully completing the test. Then take a few deep, relaxing breaths and return to it. Work through the questions carefully but with confidence, knowing that you are capable of succeeding.

Developing a healthy mental approach to test taking will also aid in other areas of life. Test anxiety affects more than just the actual test—it can be damaging to your mental health and even contribute to depression. It's important to beat test anxiety before it becomes a problem for more than testing.

Review Video: Test Anxiety and Depression
Visit mometrix.com/academy and enter code: 904704

Study Strategy

Being prepared for the test is necessary to combat anxiety, but what does being prepared look like? You may study for hours on end and still not feel prepared. What you need is a strategy for test prep. The next few pages outline our recommended steps to help you plan out and conquer the challenge of preparation.

STEP 1: SCOPE OUT THE TEST

Learn everything you can about the format (multiple choice, essay, etc.) and what will be on the test. Gather any study materials, course outlines, or sample exams that may be available. Not only will this help you to prepare, but knowing what to expect can help to alleviate test anxiety.

STEP 2: MAP OUT THE MATERIAL

Look through the textbook or study guide and make note of how many chapters or sections it has. Then divide these over the time you have. For example, if a book has 15 chapters and you have five days to study, you need to cover three chapters each day. Even better, if you have the time, leave an extra day at the end for overall review after you have gone through the material in depth.

If time is limited, you may need to prioritize the material. Look through it and make note of which sections you think you already have a good grasp on, and which need review. While you are studying, skim quickly through the familiar sections and take more time on the challenging parts. Write out your plan so you don't get lost as you go. Having a written plan also helps you feel more in control of the study, so anxiety is less likely to arise from feeling overwhelmed at the amount to cover.

STEP 3: GATHER YOUR TOOLS

Decide what study method works best for you. Do you prefer to highlight in the book as you study and then go back over the highlighted portions? Or do you type out notes of the important information? Or is it helpful to make flashcards that you can carry with you? Assemble the pens, index cards, highlighters, post-it notes, and any other materials you may need so you won't be distracted by getting up to find things while you study.

If you're having a hard time retaining the information or organizing your notes, experiment with different methods. For example, try color-coding by subject with colored pens, highlighters, or post-it notes. If you learn better by hearing, try recording yourself reading your notes so you can listen while in the car, working out, or simply sitting at your desk. Ask a friend to quiz you from your flashcards, or try teaching someone the material to solidify it in your mind.

STEP 4: CREATE YOUR ENVIRONMENT

It's important to avoid distractions while you study. This includes both the obvious distractions like visitors and the subtle distractions like an uncomfortable chair (or a too-comfortable couch that makes you want to fall asleep). Set up the best study environment possible: good lighting and a comfortable work area. If background music helps you focus, you may want to turn it on, but otherwise keep the room quiet. If you are using a computer to take notes, be sure you don't have any other windows open, especially applications like social media, games, or anything else that could distract you. Silence your phone and turn off notifications. Be sure to keep water close by so you stay hydrated while you study (but avoid unhealthy drinks and snacks).

Also, take into account the best time of day to study. Are you freshest first thing in the morning? Try to set aside some time then to work through the material. Is your mind clearer in the afternoon or evening? Schedule your study session then. Another method is to study at the same time of day that you will take the test, so that your brain gets used to working on the material at that time and will be ready to focus at test time.

Step 5: Study!

Once you have done all the study preparation, it's time to settle into the actual studying. Sit down, take a few moments to settle your mind so you can focus, and begin to follow your study plan. Don't give in to distractions or let yourself procrastinate. This is your time to prepare so you'll be ready to fearlessly approach the test. Make the most of the time and stay focused.

Of course, you don't want to burn out. If you study too long you may find that you're not retaining the information very well. Take regular study breaks. For example, taking five minutes out of every hour to walk briskly, breathing deeply and swinging your arms, can help your mind stay fresh.

As you get to the end of each chapter or section, it's a good idea to do a quick review. Remind yourself of what you learned and work on any difficult parts. When you feel that you've mastered the material, move on to the next part. At the end of your study session, briefly skim through your notes again.

But while review is helpful, cramming last minute is NOT. If at all possible, work ahead so that you won't need to fit all your study into the last day. Cramming overloads your brain with more information than it can process and retain, and your tired mind may struggle to recall even previously learned information when it is overwhelmed with last-minute study. Also, the urgent nature of cramming and the stress placed on your brain contribute to anxiety. You'll be more likely to go to the test feeling unprepared and having trouble thinking clearly.

So don't cram, and don't stay up late before the test, even just to review your notes at a leisurely pace. Your brain needs rest more than it needs to go over the information again. In fact, plan to finish your studies by noon or early afternoon the day before the test. Give your brain the rest of the day to relax or focus on other things, and get a good night's sleep. Then you will be fresh for the test and better able to recall what you've studied.

Step 6: Take a practice test

Many courses offer sample tests, either online or in the study materials. This is an excellent resource to check whether you have mastered the material, as well as to prepare for the test format and environment.

Check the test format ahead of time: the number of questions, the type (multiple choice, free response, etc.), and the time limit. Then create a plan for working through them. For example, if you have 30 minutes to take a 60-question test, your limit is 30 seconds per question. Spend less time on the questions you know well so that you can take more time on the difficult ones.

If you have time to take several practice tests, take the first one open book, with no time limit. Work through the questions at your own pace and make sure you fully understand them. Gradually work up to taking a test under test conditions: sit at a desk with all study materials put away and set a timer. Pace yourself to make sure you finish the test with time to spare and go back to check your answers if you have time.

After each test, check your answers. On the questions you missed, be sure you understand why you missed them. Did you misread the question (tests can use tricky wording)? Did you forget the information? Or was it something you hadn't learned? Go back and study any shaky areas that the practice tests reveal.

Taking these tests not only helps with your grade, but also aids in combating test anxiety. If you're already used to the test conditions, you're less likely to worry about it, and working through tests until you're scoring well gives you a confidence boost. Go through the practice tests until you feel comfortable, and then you can go into the test knowing that you're ready for it.

Test Tips

On test day, you should be confident, knowing that you've prepared well and are ready to answer the questions. But aside from preparation, there are several test day strategies you can employ to maximize your performance.

First, as stated before, get a good night's sleep the night before the test (and for several nights before that, if possible). Go into the test with a fresh, alert mind rather than staying up late to study.

Try not to change too much about your normal routine on the day of the test. It's important to eat a nutritious breakfast, but if you normally don't eat breakfast at all, consider eating just a protein bar. If you're a coffee drinker, go ahead and have your normal coffee. Just make sure you time it so that the caffeine doesn't wear off right in the middle of your test. Avoid sugary beverages, and drink enough water to stay hydrated but not so much that you need a restroom break 10 minutes into the test. If your test isn't first thing in the morning, consider going for a walk or doing a light workout before the test to get your blood flowing.

Allow yourself enough time to get ready, and leave for the test with plenty of time to spare so you won't have the anxiety of scrambling to arrive in time. Another reason to be early is to select a good seat. It's helpful to sit away from doors and windows, which can be distracting. Find a good seat, get out your supplies, and settle your mind before the test begins.

When the test begins, start by going over the instructions carefully, even if you already know what to expect. Make sure you avoid any careless mistakes by following the directions.

Then begin working through the questions, pacing yourself as you've practiced. If you're not sure on an answer, don't spend too much time on it, and don't let it shake your confidence. Either skip it and come back later, or eliminate as many wrong answers as possible and guess among the remaining ones. Don't dwell on these questions as you continue—put them out of your mind and focus on what lies ahead.

Be sure to read all of the answer choices, even if you're sure the first one is the right answer. Sometimes you'll find a better one if you keep reading. But don't second-guess yourself if you do immediately know the answer. Your gut instinct is usually right. Don't let test anxiety rob you of the information you know.

If you have time at the end of the test (and if the test format allows), go back and review your answers. Be cautious about changing any, since your first instinct tends to be correct, but make sure you didn't misread any of the questions or accidentally mark the wrong answer choice. Look over any you skipped and make an educated guess.

At the end, leave the test feeling confident. You've done your best, so don't waste time worrying about your performance or wishing you could change anything. Instead, celebrate the successful completion of this test. And finally, use this test to learn how to deal with anxiety even better next time.

> **Review Video: 5 Tips to Beat Test Anxiety**
> Visit mometrix.com/academy and enter code: 570656

Important Qualification

Not all anxiety is created equal. If your test anxiety is causing major issues in your life beyond the classroom or testing center, or if you are experiencing troubling physical symptoms related to your anxiety, it may be a sign of a serious physiological or psychological condition. If this sounds like your situation, we strongly encourage you to seek professional help.

How to Overcome Your Fear of Math

The word *math* is enough to strike fear into most hearts. How many of us have memories of sitting through confusing lectures, wrestling over mind-numbing homework, or taking tests that still seem incomprehensible even after hours of study? Years after graduation, many still shudder at these memories.

The fact is, math is not just a classroom subject. It has real-world implications that you face every day, whether you realize it or not. This may be balancing your monthly budget, deciding how many supplies to buy for a project, or simply splitting a meal check with friends. The idea of daily confrontations with math can be so paralyzing that some develop a condition known as *math anxiety*.

But you do NOT need to be paralyzed by this anxiety! In fact, while you may have thought all your life that you're not good at math, or that your brain isn't wired to understand it, the truth is that you may have been conditioned to think this way. From your earliest school days, the way you were taught affected the way you viewed different subjects. And the way math has been taught has changed.

Several decades ago, there was a shift in American math classrooms. The focus changed from traditional problem-solving to a conceptual view of topics, de-emphasizing the importance of learning the basics and building on them. The solid foundation necessary for math progression and confidence was undermined. Math became more of a vague concept than a concrete idea. Today, it is common to think of math, not as a straightforward system, but as a mysterious, complicated method that can't be fully understood unless you're a genius.

This is why you may still have nightmares about being called on to answer a difficult problem in front of the class. Math anxiety is a very real, though unnecessary, fear.

Math anxiety may begin with a single class period. Let's say you missed a day in 6th grade math and never quite understood the concept that was taught while you were gone. Since math is cumulative, with each new concept building on past ones, this could very well affect the rest of your math career. Without that one day's knowledge, it will be difficult to understand any other concepts that link to it. Rather than realizing that you're just missing one key piece, you may begin to believe that you're simply not capable of understanding math.

This belief can change the way you approach other classes, career options, and everyday life experiences, if you become anxious at the thought that math might be required. A student who loves science may choose a different path of study upon realizing that multiple math classes will be required for a degree. An aspiring medical student may hesitate at the thought of going through the necessary math classes. For some this anxiety escalates into a more extreme state known as *math phobia*.

Math anxiety is challenging to address because it is rooted deeply and may come from a variety of causes: an embarrassing moment in class, a teacher who did not explain concepts well and contributed to a shaky foundation, or a failed test that contributed to the belief of math failure.

These causes add up over time, encouraged by society's popular view that math is hard and unpleasant. Eventually a person comes to firmly believe that he or she is simply bad at math. This belief makes it difficult to grasp new concepts or even remember old ones. Homework and test grades begin to slip, which only confirms the belief. The poor performance is not due to lack of ability but is caused by math anxiety.

Math anxiety is an emotional issue, not a lack of intelligence. But when it becomes deeply rooted, it can become more than just an emotional problem. Physical symptoms appear. Blood pressure may rise and heartbeat may quicken at the sight of a math problem – or even the thought of math! This fear leads to a mental block. When someone with math anxiety is asked to perform a calculation, even a basic problem can seem overwhelming and impossible. The emotional and physical response to the thought of math prevents the brain from working through it logically.

The more this happens, the more a person's confidence drops, and the more math anxiety is generated. This vicious cycle must be broken!

The first step in breaking the cycle is to go back to very beginning and make sure you really understand the basics of how math works and why it works. It is not enough to memorize rules for multiplication and division. If you don't know WHY these rules work, your foundation will be shaky and you will be at risk of developing a phobia. Understanding mathematical concepts not only promotes confidence and security, but allows you to build on this understanding for new concepts. Additionally, you can solve unfamiliar problems using familiar concepts and processes.

Why is it that students in other countries regularly outperform American students in math? The answer likely boils down to a couple of things: the foundation of mathematical conceptual understanding and societal perception. While students in the US are not expected to *like* or *get* math, in many other nations, students are expected not only to understand math but also to excel at it.

Changing the American view of math that leads to math anxiety is a monumental task. It requires changing the training of teachers nationwide, from kindergarten through high school, so that they learn to teach the *why* behind math and to combat the wrong math views that students may develop. It also involves changing the stigma associated with math, so that it is no longer viewed as unpleasant and incomprehensible. While these are necessary changes, they are challenging and will take time. But in the meantime, math anxiety is not irreversible—it can be faced and defeated, one person at a time.

False Beliefs

One reason math anxiety has taken such hold is that several false beliefs have been created and shared until they became widely accepted. Some of these unhelpful beliefs include the following:

There is only one way to solve a math problem. In the same way that you can choose from different driving routes and still arrive at the same house, you can solve a math problem using different methods and still find the correct answer. A person who understands the reasoning behind math calculations may be able to look at an unfamiliar concept and find the right answer, just by applying logic to the knowledge they already have. This approach may be different than what is taught in the classroom, but it is still valid. Unfortunately, even many teachers view math as a subject where the best course of action is to memorize the rule or process for each problem rather than as a place for students to exercise logic and creativity in finding a solution.

Many people don't have a mind for math. A person who has struggled due to poor teaching or math anxiety may falsely believe that he or she doesn't have the mental capacity to grasp mathematical concepts. Most of the time, this is false. Many people find that when they are relieved of their math anxiety, they have more than enough brainpower to understand math.

Men are naturally better at math than women. Even though research has shown this to be false, many young women still avoid math careers and classes because of their belief that their math abilities are inferior. Many girls have come to believe that math is a male skill and have given up trying to understand or enjoy it.

Counting aids are bad. Something like counting on your fingers or drawing out a problem to visualize it may be frowned on as childish or a crutch, but these devices can help you get a tangible understanding of a problem or a concept.

Sadly, many students buy into these ideologies at an early age. A young girl who enjoys math class may be conditioned to think that she doesn't actually have the brain for it because math is for boys, and may turn her energies to other pursuits, permanently closing the door on a wide range of opportunities. A child who finds the right answer but doesn't follow the teacher's method may believe that he is doing it wrong and isn't good at math. A student who never had a problem with math before may have a poor teacher and become confused, yet believe that the problem is because she doesn't have a mathematical mind.

Students who have bought into these erroneous beliefs quickly begin to add their own anxieties, adapting them to their own personal situations:

I'll never use this in real life. A huge number of people wrongly believe that math is irrelevant outside the classroom. By adopting this mindset, they are handicapping themselves for a life in a mathematical world, as well as limiting their career choices. When they are inevitably faced with real-world math, they are conditioning themselves to respond with anxiety.

I'm not quick enough. While timed tests and quizzes, or even simply comparing yourself with other students in the class, can lead to this belief, speed is not an indicator of skill level. A person can work very slowly yet understand at a deep level.

If I can understand it, it's too easy. People with a low view of their own abilities tend to think that if they are able to grasp a concept, it must be simple. They cannot accept the idea that they are capable of understanding math. This belief will make it harder to learn, no matter how intelligent they are.

I just can't learn this. An overwhelming number of people think this, from young children to adults, and much of the time it is simply not true. But this mindset can turn into a self-fulfilling prophecy that keeps you from exercising and growing your math ability.

The good news is, each of these myths can be debunked. For most people, they are based on emotion and psychology, NOT on actual ability! It will take time, effort, and the desire to change, but change is possible. Even if you have spent years thinking that you don't have the capability to understand math, it is not too late to uncover your true ability and find relief from the anxiety that surrounds math.

Math Strategies

It is important to have a plan of attack to combat math anxiety. There are many useful strategies for pinpointing the fears or myths and eradicating them:

Go back to the basics. For most people, math anxiety stems from a poor foundation. You may think that you have a complete understanding of addition and subtraction, or even decimals and percentages, but make absolutely sure. Learning math is different from learning other subjects. For example, when you learn history, you study various time periods and places and events. It may be important to memorize dates or find out about the lives of famous people. When you move from US history to world history, there will be some overlap, but a large amount of the information will be new. Mathematical concepts, on the other hand, are very closely linked and highly dependent on each other. It's like climbing a ladder – if a rung is missing from your understanding, it may be difficult or impossible for you to climb any higher, no matter how hard you try. So go back and make sure your math foundation is strong. This may mean taking a remedial math course, going to a tutor to work through the shaky concepts, or just going through your old homework to make sure you really understand it.

Speak the language. Math has a large vocabulary of terms and phrases unique to working problems. Sometimes these are completely new terms, and sometimes they are common words, but are used differently in a math setting. If you can't speak the language, it will be very difficult to get a thorough understanding of the concepts. It's common for students to think that they don't understand math when they simply don't understand the vocabulary. The good news is that this is fairly easy to fix. Brushing up on any terms you aren't quite sure of can help bring the rest of the concepts into focus.

Check your anxiety level. When you think about math, do you feel nervous or uncomfortable? Do you struggle with feelings of inadequacy, even on concepts that you know you've already learned? It's important to understand your specific math anxieties, and what triggers them. When you catch yourself falling back on a false belief, mentally replace it with the truth. Don't let yourself believe that you can't learn, or that struggling with a concept means you'll never understand it. Instead, remind yourself of how much you've already learned and dwell on that past success. Visualize grasping the new concept, linking it to your old knowledge, and moving on to the next challenge. Also, learn how to manage anxiety when it arises. There are many techniques for coping with the irrational fears that rise to the surface when you enter the math classroom. This may include controlled breathing, replacing negative thoughts with positive ones, or visualizing success. Anxiety interferes with your ability to concentrate and absorb information, which in turn contributes to greater anxiety. If you can learn how to regain control of your thinking, you will be better able to pay attention, make progress, and succeed!

Don't go it alone. Like any deeply ingrained belief, math anxiety is not easy to eradicate. And there is no need for you to wrestle through it on your own. It will take time, and many people find that speaking with a counselor or psychiatrist helps. They can help you develop strategies for responding to anxiety and overcoming old ideas. Additionally, it can be very helpful to take a short course or seek out a math tutor to help you find and fix the missing rungs on your ladder and make sure that you're ready to progress to the next level. You can also find a number of math aids online: courses that will teach you mental devices for figuring out problems, how to get the most out of your math classes, etc.

Check your math attitude. No matter how much you want to learn and overcome your anxiety, you'll have trouble if you still have a negative attitude toward math. If you think it's too hard, or just have general feelings of dread about math, it will be hard to learn and to break through the anxiety. Work on cultivating a positive math attitude. Remind yourself that math is not just a hurdle to be cleared, but a valuable asset. When you view math with a positive attitude, you'll be much more likely to understand and even enjoy it. This is something you must do for yourself. You may find it helpful to visit with a counselor. Your tutor, friends, and family may cheer you on in your endeavors. But your greatest asset is yourself. You are inside your own mind – tell yourself what you need to hear. Relive past victories. Remind yourself that you are capable of understanding math. Root out any false beliefs that linger and replace them with positive truths. Even if it doesn't feel true at first, it will begin to affect your thinking and pave the way for a positive, anxiety-free mindset.

Aside from these general strategies, there are a number of specific practical things you can do to begin your journey toward overcoming math anxiety. Something as simple as learning a new note-taking strategy can change the way you approach math and give you more confidence and understanding. New study techniques can also make a huge difference.

Math anxiety leads to bad habits. If it causes you to be afraid of answering a question in class, you may gravitate toward the back row. You may be embarrassed to ask for help. And you may procrastinate on assignments, which leads to rushing through them at the last moment when it's too late to get a better understanding. It's important to identify your negative behaviors and replace them with positive ones:

Prepare ahead of time. Read the lesson before you go to class. Being exposed to the topics that will be covered in class ahead of time, even if you don't understand them perfectly, is extremely helpful in increasing what you retain from the lecture. Do your homework and, if you're still shaky, go over some extra problems. The key to a solid understanding of math is practice.

Sit front and center. When you can easily see and hear, you'll understand more, and you'll avoid the distractions of other students if no one is in front of you. Plus, you're more likely to be sitting with students who are positive and engaged, rather than others with math anxiety. Let their positive math attitude rub off on you.

Ask questions in class and out. If you don't understand something, just ask. If you need a more in-depth explanation, the teacher may need to work with you outside of class, but often it's a simple concept you don't quite understand, and a single question may clear it up. If you wait, you may not be able to follow the rest of the day's lesson. For extra help, most professors have office hours outside of class when you can go over concepts one-on-one to clear up any uncertainties. Additionally, there may be a *math lab* or study session you can attend for homework help. Take advantage of this.

Review. Even if you feel that you've fully mastered a concept, review it periodically to reinforce it. Going over an old lesson has several benefits: solidifying your understanding, giving you a confidence boost, and even giving some new insights into material that you're currently learning! Don't let yourself get rusty. That can lead to problems with learning later concepts.

Teaching Tips

While the math student's mindset is the most crucial to overcoming math anxiety, it is also important for others to adjust their math attitudes. Teachers and parents have an enormous influence on how students relate to math. They can either contribute to math confidence or math anxiety.

As a parent or teacher, it is very important to convey a positive math attitude. Retelling horror stories of your own bad experience with math will contribute to a new generation of math anxiety. Even if you don't share your experiences, others will be able to sense your fears and may begin to believe them.

Even a careless comment can have a big impact, so watch for phrases like *He's not good at math* or *I never liked math*. You are a crucial role model, and your children or students will unconsciously adopt your mindset. Give them a positive example to follow. Rather than teaching them to fear the math world before they even know it, teach them about all its potential and excitement.

Work to present math as an integral, beautiful, and understandable part of life. Encourage creativity in solving problems. Watch for false beliefs and dispel them. Cross the lines between subjects: integrate history, English, and music with math. Show students how math is used every day, and how the entire world is based on mathematical principles, from the pull of gravity to the shape of seashells. Instead of letting students see math as a necessary evil, direct them to view it as an imaginative, beautiful art form – an art form that they are capable of mastering and using.

Don't give too narrow a view of math. It is more than just numbers. Yes, working problems and learning formulas is a large part of classroom math. But don't let the teaching stop there. Teach students about the everyday implications of math. Show them how nature works according to the laws of mathematics, and take them outside to make discoveries of their own. Expose them to math-related careers by inviting visiting speakers, asking students to do research and presentations, and learning students' interests and aptitudes on a personal level.

Demonstrate the importance of math. Many people see math as nothing more than a required stepping stone to their degree, a nuisance with no real usefulness. Teach students that algebra is used every day in managing their bank accounts, in following recipes, and in scheduling the day's events. Show them how learning to do geometric proofs helps them to develop logical thinking, an invaluable life skill. Let them see that math surrounds them and is integrally linked to their daily lives: that weather predictions are based on math, that math was used to design cars and other machines, etc. Most of all, give them the tools to use math to enrich their lives.

Make math as tangible as possible. Use visual aids and objects that can be touched. It is much easier to grasp a concept when you can hold it in your hands and manipulate it, rather than just listening to the lecture. Encourage math outside of the classroom. The real world is full of measuring, counting, and calculating, so let students participate in this. Keep your eyes open for numbers and patterns to discuss. Talk about how scores are calculated in sports games and how far apart plants are placed in a garden row for maximum growth. Build the mindset that math is a normal and interesting part of daily life.

Finally, find math resources that help to build a positive math attitude. There are a number of books that show math as fascinating and exciting while teaching important concepts, for example: *The Math Curse*; *A Wrinkle in Time*; *The Phantom Tollbooth*; and *Fractals, Googols and Other Mathematical Tales*. You can also find a number of online resources: math puzzles and games, videos that show math in nature, and communities of math enthusiasts. On a local level, students can compete in a variety of math competitions with other schools or join a math club.

The student who experiences math as exciting and interesting is unlikely to suffer from math anxiety. Going through life without this handicap is an immense advantage and opens many doors that others have closed through their fear.

Self-Check

Whether you suffer from math anxiety or not, chances are that you have been exposed to some of the false beliefs mentioned above. Now is the time to check yourself for any errors you may have accepted. Do you think you're not wired for math? Or that you don't need to understand it since you're not planning on a math career? Do you think math is just too difficult for the average person?

Find the errors you've taken to heart and replace them with positive thinking. Are you capable of learning math? Yes! Can you control your anxiety? Yes! These errors will resurface from time to time, so be watchful. Don't let others with math anxiety influence you or sway your confidence. If you're having trouble with a concept, find help. Don't let it discourage you!

Create a plan of attack for defeating math anxiety and sharpening your skills. Do some research and decide if it would help you to take a class, get a tutor, or find some online resources to fine-tune your knowledge. Make the effort to get good nutrition, hydration, and sleep so that you are operating at full capacity. Remind yourself daily that you are skilled and that anxiety does not control you. Your mind is capable of so much more than you know. Give it the tools it needs to grow and thrive.

Thank You

We at Mometrix would like to extend our heartfelt thanks to you, our friend and patron, for allowing us to play a part in your journey. It is a privilege to serve people from all walks of life who are unified in their commitment to building the best future they can for themselves.

The preparation you devote to these important testing milestones may be the most valuable educational opportunity you have for making a real difference in your life. We encourage you to put your heart into it—that feeling of succeeding, overcoming, and yes, conquering will be well worth the hours you've invested.

We want to hear your story, your struggles and your successes, and if you see any opportunities for us to improve our materials so we can help others even more effectively in the future, please share that with us as well. **The team at Mometrix would be absolutely thrilled to hear from you!** So please, send us an email (support@mometrix.com) and let's stay in touch.

> **If you'd like some additional help, check out these other resources we offer for your exam:**
> **http://MometrixFlashcards.com/CHST**

Additional Bonus Material

Due to our efforts to try to keep this book to a manageable length, we've created a link that will give you access to all of your additional bonus material.

> Please visit https://www.mometrix.com/bonus948/chst to access the information.

Made in the USA
Coppell, TX
24 October 2019